FOOD SCIENCE AND TECHNOLOGY

FOOD SCIENCE RESEARCH SUMMARIES

VOLUME 3

FOOD SCIENCE AND TECHNOLOGY

Additional books in this series can be found on Nova's website
under the Series tab.

Additional e-books in this series can be found on Nova's website
under the e-book tab.

FOOD SCIENCE AND TECHNOLOGY

FOOD SCIENCE RESEARCH SUMMARIES

VOLUME 3

LUCILLE MONACO CACIOPPO
EDITOR

nova publishers
New York

For permission to use material from this book please contact us:
Telephone 631-231-7269; Fax 631-231-8175
Web Site: http://www.novapublishers.com

NOTICE TO THE READER

Additional color graphics may be available in the e-book version of this book.

Library of Congress Cataloging-in-Publication Data

ISBN: 978-1-63117-864-1

Published by Nova Science Publishers, Inc. † New York

CONTENTS

Preface xvii

Chapter 1 Technological Evaluation of Pan Bread Produced with the Addition of Commercial Resistant Starch or Unripe Banana Flour 1
Caroline Joy Steel, Camila Batista da Silva, Rita de Cássia Salvucci Celeste Ormenese, Maria Teresa Pedrosa Silva Clerici, Miluska Castilho Cáceres and Yoon Kil Chang

Chapter 2 Importance of French-type Bread in Brazilian Diet 3
Maria Teresa Pedrosa Silva Clerici, Elizabeth Harumi Nabeshima, Jorge Minoru Hashimoto, Luciano Bruno de Carvalho-Silva and Yoon Kil Chang

Chapter 3 Effect of Adding Ascorbic Acid on the Quality of Sliced Loaf with Partial Substitution of Amaranth Flour (Amaranthus Caudatus) and Wheat Bran (Triticum Aestivum) 5
Luz Maria Paucar-Menacho, Nilthon Santa María-Podestá and Nathaly Aguedita Cerna-Lucio

Chapter 4 Maize Bread: Healthy and Safe 7
Malvar Rosa Ana, Butrón Ana and Revilla Pedro

Chapter 5 Nutritional Quality of Proteins in Breads 9
Luciano Bruno de Carvalho-Silva and Maria Teresa Pedrosa Silva Clerici

Chapter 6 Fiber-Enriched Bread as an Aid in the Nutritional Management of Dislipemic Patients 11
Ismael Darío Bianco, Dante Miguel Beltramo and Gustavo Tomás Diaz

Chapter 7	Effect of Different Pineapple Juice (*Ananas comosus* l.) Preparations on the Microstructure, Staling and Textural Properties of Wheat Bread *Ola Lasekan, Chinma Chiemela, Beatrice Ossai and Noranizan Mohd Adzahan*	**13**
Chapter 8	Nutritional and Technological Characteristics of Gluten-Free and Low Glycemic Index Baking Products *A. M. Calderón de la Barca, M. I. Ortega-Vélez and F. Cabrera-Chávez*	**15**
Chapter 9	Experimental Design for the Evaluation of Bread Protein Quality *Luciano Bruno de Carvalho-Silva and Maria Teresa Pedrosa Silva Clerici*	**17**
Chapter 10	Packaging and Required Protection for the Preservation of Bread Functional Compounds *Eduardo Henrique Miranda Walter*	**19**
Chapter 11	Gastrointestinal Health Promotion via Bread Consumption: Elucidation of Preventive Effects in the Etiology of Pathological Events along the Intestine *Beate Hiller' Wiebke Schlörmann, Jürgen Hollmann, Michael Glei and Meinolf Lindhauer*	**21**
Chapter 12	What Is Coffee? *Bernard A. Goodman*	**23**
Chapter 13	Antioxidant and Free Radical Processes in Coffee *Bernard A. Goodman*	**25**
Chapter 14	Properties and Analysis of Caffeine and Chlorogenic Acids in Coffee Beans *Abebe Belay*	**27**
Chapter 15	Occurrence and Identification of Chlorogenic Acids in Green and Roasted Coffee Beans *Nikolai Kuhnert and Rakesh Jaiswal*	**29**
Chapter 16	Coffee Roasting: Accurate Control for Increased Bioactivity *Susana Casal*	**31**
Chapter 17	Characterization of Organic Coffee Using Photothermal Techniques *F. Gordillo-Delgado and E. Marín*	**33**
Chapter 18	Coffee Consumption and the Health of the Gastrointestinal and Endocrine Systems *Bernard A. Goodman*	**35**
Chapter 19	Effects of Coffee on Inflammatory Processes *Bernard A. Goodman*	**37**

Chapter 20	Epidemiological Evidence for the Association of Coffee Consumption with Cancer *Bernard A. Goodman*	**39**
Chapter 21	Coffee Consumption and Oxidative DNA Damage *Takeshi Hirano', Haruki Morii and Kazuyoshi Tamae*	**41**
Chapter 22	Coffee, Caffeine and the Cardiovascular System *Bernard A. Goodman*	**43**
Chapter 23	Coffee Consumption and Type 2 Diabetes Mellitus *Juliane Costa Silva Zemdegs, Cristiane de Oliveira, Anita Sachs and João Felipe Mota*	**45**
Chapter 24	Neurological Effects of Coffee Consumption *Bernard A. Goodman*	**47**
Chapter 25	Effects of Coffee Consumption during Pregnancy and in Young Children *Bernard A. Goodman*	**49**
Chapter 26	Effects of Caffeine Consumption on Exercise *Bernard A. Goodman*	**51**
Chapter 27	Other Health-Related Aspects of Coffee Consumption *Bernard A. Goodman*	**53**
Chapter 28	Effect of Ionizing Radiation on Polymeric Food Packaging Materials *A. Badeka and M. G. Kontominas*	**55**
Chapter 29	Packaging Optimization for Minimally Processed Fruit and Vegetables *Amalia Conte, Marcella Mastromatteo and Matteo Alessandro Del Nobile*	**57**
Chapter 30	Characterization of PLA, PCL and Sodium Caseinate Active Bio-Films for Food Packaging Applications *M. Ramos, M. P. Arrieta, A. Beltran and M. C. Garrigós*	**59**
Chapter 31	Metal-Based Micro and Nano-Composites as Antimicrobials in Food Packaging *Avelina Fernandez, Elsa Lloret, Amparo Llorens and Pierre Picouet*	**61**
Chapter 32	Alternative Technique for the Quantitative Determination of Antimicrobial Activity of Antimicrobial Packaging Film Containing Lipophilic Materials *Panuwat Suppakul*	**63**
Chapter 33	Antimicrobial Packaging: Lab Developments and Possible Commercialization *Kay Cooksey, Angela Richard and Sunil Mangalassary*	**65**

Chapter 34	Evaluation of Safety of Recycled Paper and Board Food Contact Materials *Panagiotis G. Demertzis*	**67**
Chapter 35	Faking Authenticity? *Maria Kniazeva*	**69**
Chapter 36	The Multisensory Packaging of Beverages *Charles Spence and Betina Piqueras-Fiszman*	**71**
Chapter 37	Water Soluble Components in Green Teas and Their Effects on Human Health *Wenbiao Wu and Mingliang Huang*	**73**
Chapter 38	Green Tea: Varieties, Production and Health Benefits *Nesreen Ali, Eman Roshdy, Mohamed Sabry and Ayman Al-Hendy*	**75**
Chapter 39	Green Tea and Pancreatic Cancer Chemoprevention *Quan V. Vuong, Christopher J. Scarlett and Paul D. Roach*	**77**
Chapter 40	Effects of Green Tea on Cognitive Function and Mental Status: Emerging Evidence *Wenbin Liang and Andy H. Lee*	**79**
Chapter 41	Effects of Anti-Oxidative Stress of Green Tea Catechins on Cardiovascular Diseases *Jun-ichi Suzuki, Mitsuaki Isobe, Yasunobu Hirata and Ryozo Nagai*	**81**
Chapter 42	Antidiabetic Properties of Green Tea Catechins *Yannis V. Simos, Evangelia Tselikou, Katerina Touloupi, Ioannis Verginadis, Ilias Karagounis, Anastasia Velalopoulou and Angelos Evangelou*	**83**
Chapter 43	Polyphenol Tea: Protein Protection, Cancer Prevention and Nanotechnology *Kouass Sahbani Saloua, Kouass Salah, Aloui Amine, Abbas Chiraz, El May Alya and Landoulsi Ahmed*	**85**
Chapter 44	Green Tea and Skin Benefits *Neti Waranuch and Wudtichai Wisuitiprot*	**87**
Chapter 45	Non-covalent Interactions of Green Tea Polyphenol: Source of Novel Colloidal Structures *Ashok R. Patel and Krassimir P. Velikov*	**89**
Chapter 46	Chinese Green Tea: Production, Varietal Resources and Adaptability *La Li and Chun Xiao*	**91**

Chapter 47 Phytochemicals as Potential Pharmacophores to Target Diabetes **93**
Mahomoodally Mohamad Fawzi

Chapter 48 Phytochemicals and their Potential Effects on Inflammatory Bowel Disease **95**
Susana Gorzalczany

Chapter 49 Bioactive Compounds in Whole Grain Cereals **97**
Maria Teresa Pedrosa Silva Clerici and Elizabeth Harumi Nabeshima

Chapter 50 Antioxidant and Antimicrobial Activities of Berries Extracts of Five *Juniperus* Species under *Juniperus* Section from Turkey **99**
Maria Fernanda Taviano, Andreana Marino, Ada Trovato, Rita De Pasquale and Natalizia Miceli

Chapter 51 Phenolic Compounds, Health Effects and Antioxidant Properties of Berries **101**
Aleksandra Konić-Ristić, Katarina Šavikin, Gordana Zdunić, Dejan Gođevac, Teodora Janković and Nebojša Menković

Chapter 52 Overview of the Chemical, Natural Occurrence and Antioxidant Activity of Terpenes **103**
E. González-Burgos and M. P. Gómez-Serranillos

Chapter 53 Supercritical Extraction of Phloroglucinols from St. John's Wort (*Hypericum Perforatum* L.) – Process Optimization, Mathematical Modeling and Antimicrobial Activity **105**
Sandra B. Glisic and Aleksandar M. Orlovic

Chapter 54 Natural Antioxidants in Food **107**
Gui-Fang Deng, Xiang-Rong Xu, Sha Li, Fang Li, Shan Wu and Hua-Bin Li

Chapter 55 Regulation of Inflammation by Dietary Phytochemicals: Current Perspectives **109**
Marta González Castejón and Arantxa Rodriguez Casado

Chapter 56 Antioxidative Phytochemicals and Skin Care **111**
Frédéric Bonté and Gallic Beauchef

Chapter 57 Research Progress on Medicinal Plants of Urtica Linn. Genus **113**
Zhang Yanxiu, Wang Ying, Ju Zhihe and Guan Feng

Chapter 58 Research and Development on Chemical Constituents and Pharmacology of *Acanthopanax* Genus **115**
Gaofeng Liu, Chao Zeng and Zhiping Lin

Chapter 59 Simultaneous Determination of Five Triterpenoid Saponins in the Root of *Caulophyllum Robustum* Maxim by HPLC-ELSD **117**
Guo-Yu Li, Yong-Gang Xia, Jun Liang, Yan-Hai Zhang and Hai-Xue Kuang

Chapter 60 Chemical Study and Medical Application of Saponins from the Genus *Aralia* **119**
Hai-Xue Kuang, Yong-Gang Xia and Guang-Rui Mu

Chapter 61 Chemical Constituents and Bioactivities of Plants of the Genus *Hydrangea* **121**
Yang Chunjuan and Xiao Yang*

Chapter 62 Phytochemicals in Cancer Research: A Review of Their Biological Activities and Anticancer Potentials in Preclinical and Clinical Studies **123**
Daniel L. Pouliquen and Marc Grégoire

Chapter 63 Mammary Glands, Aquaporins and Milk Production **125**
Ali Mobasheri

Chapter 64 Science and Pseudo-Science of Milk Implications for Human Health **127**
A. Nikkhah

Chapter 65 Ruminant Milk and Human Wellbeing: A Multi-Species Review of the Natural Relations **129**
A. Nikkhah

Chapter 66 Colostral and Lactogenic Maternal Immunity: Humoral and Cellular Factors of Induction and Transmission to the Neonate **131**
H. Salmon

Chapter 67 The Importance of Milk Fatty Acids in Human Nutrition **133**
Rey Gutiérrez Tolentino, Salvador Vega y León, Claudia Radilla Vázquez, María Radilla Vázquez, Samuel Coronel Nuñez and Marta Coronado Herrera

Chapter 68 Investigations on the Geographical Origin of Cow Milk and Comparison of the Lipid Composition of Cow and Buffalo Milk by Means of Traditional and Innovative Physico-Chemical Analyses **135**
A. Sacco, D. Sacco, G. Casiello, V. Mazzilli, A. Ventrella and F. Longobardi

Chapter 69 Alternative Approaches for the Prevention of Bovine Mastitis. Probiotics, Bioactive Compounds and Vaccines **137**
María Elena Fátima Nader-Macías, Cristina Bogni, Fernando Juan Manuel Sesma, María Carolina Espeche, Matías Pellegrino, Lucila Saavedra and Ignacio Frola

Chapter 70 Milk Production of Holsteins under Mediterranean Conditions: Case of the Tunisian Population **139**
A. Ben Gara, B. Jemmali, H. Hammami, H. Rouissi, M. Bouallegue and B. Rekik

Chapter 71 Fresh Ewe Milk Production and Cereal Nutrition: A Peripartal Interactive Model of Grain Choice and Level **141**
A. Nikkhah and M. Karam-Babaei

Chapter 72 Fatty Acid Composition of Milk Lipids in Response to Dietary Fish and Safflower Oils **143**
K. L. Jacobsen, S. Harris, E. J. DePeters and S. J. Taylor

Chapter 73 Genetic Factors and Dairy-Technological Valorisation of Milk **145**
Andrea Summer, Massimo Malacarne, Paolo Formaggioni, Piero Franceschi and Primo Mariani

Chapter 74 Animal Welfare in Dairy Operations: A Postmodern View **147**
Zenobia C. Y. Chan and Wing-Fu Lai

Chapter 75 Soy: Nutrition, Consumption and Heart Health **149**
Janice I. Harland, Elaine S. Krul, Ratna Mukherjea and Kevin Maki

Chapter 76 Soy Isoflavones in Bone Health and Function **151**
Alessandra Bitto, Bruce P. Burnett and Francesco Squadrito

Chapter 77 Soy's Potential Role in Weight Management **153**
Mark B. Cope, Ratna Mukherjea, Elaine Krul and David B. Allison

Chapter 78 Antihypertensive Activity of Soy and Soy-Derived Compounds **155**
Kathleen M. Eyster, Nikolai P. Breitkopf and Douglas S. Martin

Chapter 79 Mechanisms of Anti-Atherosclerotic Activity of Soy and Soy-Derived Compounds **157**
Shanmugam Nagarajan

Chapter 80 Evaluation of Efficacy of Soy Isoflavones in Reducing the Menopausal Symptoms **159**
Márcia Constantino Colli, Cinthia Gandolfi Boer, Andréa Assunção Soares, Niladê Neves Rosinski Rocha, Adelar Bracht and Rosane Marina Peralta

Chapter 81 Soy-Derived Compounds and Immune Response **161**
Tohru Sakai, Emi Shuto and Rie Tsutsumi

Chapter 82 The Role of Non-isoflavone Phytochemicals in Soy-derived Health Benefits **163**
Xianli Wu, Jie Kang and Keshari M. Thakali

Chapter 83	Effect of Dietary Fermented Soybean on Physical Endurance in Mice *Nakamichi Watanabe and Satomi Sakuda*	**165**
Chapter 84	Soy Consumption and Self-Reported Health *Eef Hogervorst, Sumoryono, Ninu, Dwi Astuti, Sigit Prastowo, Amina Yesufu, Linda Kusdhany, Henry Setyawan, Dewi Praidini, Fidiansyah Murshid, Raden Ira Ismail, Philip Kreager, Xi Xin, Jen Stock, Angela Clifford, Stephan Bandelow, Sabarinah and Tri Budi Rahardjo*	**167**
Chapter 85	Soy Isoflavones: Bioorganic and Medicinal Chemistry of a Potent Compound Class *Sebastian Seibt*	**169**
Chapter 86	Soy Plant Tissue Culture as an Alternative for Isoflavonoid Production *A. Gueven and Z. Hicsasmaz*	**171**
Chapter 87	Removal of Antinutritional Factors in Soy-Based Foods Using Lactic Acid Bacteria *Jean Guy LeBlanc, Marisa Garro, Graciela Rollan and Graciela Savoy de Giori*	**173**
Chapter 88	Soy Ingredient Addition to Nutritionally Fortified Baked Goods *P. Roccia, P. D. Ribotta, G. T. Pérez and A. E. León*	**175**
Chapter 89	Increasing the Bioavailability of Soy Isoflavones Using Lactic Acid Bacteria *Graciela Savoy de Giori, Jose Marazza, Jean Guy LeBlanc and Marisa Garro*	**177**
Chapter 90	Combustion of Biodiesel Fuel Made from Soybean Oil *Charles K. Westbrook*	**179**
Chapter 91	Soy Isoflavone Genistein in Cancer Chemoprevention: Role of Endogenous Copper and Involvement of Reactive Oxygen Species *Haseeb Zubair, Mohammad F. Ullah, Husain Y. Khan, Fazlul H. Sarkar, Aamir Ahmad, and Sheikh M. Hadi*	**181**
Chapter 92	Chemopreventive Role of Soybean Peptides against Cancer *Chia-Chien Hsieh, Blanca Hernández-Ledesma and Ben O. de Lumen*	**183**
Chapter 93	Impact of Soyfood Intake on Breast and Endometrial Cancer Risk and Thyroid Function: Evaluating Safety Concerns *Mark Messina*	**185**

Chapter 94 A Prospective Look at Genistein Use in Cancer Prevention and Treatment **187**
Amanda C. Sehmer, Aamir Ahmad, Yiwei Li, Bin Bao and Fazlul H. Sarkar

Chapter 95 Metal and Polycyclic Aromatic Hydrocarbon Content of Products in Sugar Production **189**
Zita Šereš, Biljana Škrbić, Dragana Šoronja Simović, Biljana Pajin, Ljubica Dokić and Djordje Tadić

Chapter 96 Bioprotective Effects of Sucrose and Trehalose on Proteins **191**
Salvatore Magazù, Federica Migliardo, Antonio Benedetto, Emanuele Calabrò, Roberto La Torre and Maria Teresa Caccamo

Chapter 97 Influence of Sucrose on Surrounding Water by Extended Frequency Range Depolarized Light Scattering **193**
L. Lupi, L. Comez, M. Paolantoni, P. Sassi, A. Morresi and D. Fioretto

Chapter 98 Improvement of Color and Functional Stability of Strawberry Jam by Impregnation with Sucrose **195**
Yoshiyuki Watanabe

Chapter 99 The Effects of Sucrose on Neuronal Activity **197**
Elena Timofeeva and Arojit Mitra

Chapter 100 Effect of the Addition of Sucrose in Rheological Properties **199**
Priscila Anadão

Chapter 101 Reactions of Sucrose in the Presence of Solid Catalysts **201**
D. S. Pito and J. E. Castanheiro

Chapter 102 Cytotoxic Potential of Sucrose Octaacetate from the Ethanol Extract of the Seeds of *Annona Cornifolia* A. St.-Hil. (Annonaceae) **203**
Luciana A. R. S. Lima, Tânia M. A. Alves, Carlos L. Zani, Lúcia P. S. Pimenta and Maria Amélia D. Boaventura

Chapter 103 Myths and Facts on Cytosolic Sucrose Hydrolysis **205**
Walter A. Vargas, Mariana L. Martin and Graciela L. Salerno

Chapter 104 Tea Polyphenols: Health, Bioavailability and Novel Use **207**
Kouass Sahbani Saloua, Kouass Salah and Landoulsi Ahmed

Chapter 105 Green Tea and its Impact on Human Health: Myth or Scientific Truth? **209**
Branca M. Silva

Chapter 106 Potential Health Benefits of Green Tea: Role of Epigallocatechin-3-Gallate in Prevention of Inflammatory Pathologies **211**
Esther Melgarejo, Miguel Ángel Medina, Francisca Sánchez-Jiménez and José Luis Urdiales

Chapter 107 Effects of Green Tea Catechin Consumption on Cardiovascular Diseases **213**
Jun-ichi Suzuki, Mitsuaki Isobe, Ryuichi Morishita and Ryozo Nagai

Chapter 108 Anti-diabetic Activity of Black, Green and Oolong Tea **215**
Hardik Gandhi and Aman Upaganlawar

Chapter 109 Mechanisms Underlying Ischemic Stroke Prevention by Tea (*Camellia Sinensis*) **217**
Yuri N. Clement

Chapter 110 Green Tea Polyphenolic Compounds and Cardiovascular Disorders **219**
Aman Upaganlawar and Hardik Gandhi

Chapter 111 Tea and Healthy Skin **221**
Neti Waranuch and Wudtichai Wisuitiprot

Chapter 112 Epigallocatechin-3-gallate (EGCG), a Constituent of Green Tea, and its Antifibrotic Effect **223**
O. A. Branford, A. O. Grobbelaar and K. J. Rolfe

Chapter 113 Translational Research of the Green Tea Catechin, Epigallocatechin Gallate, as an HIV-1 Therapy **225**
Christina L. Nance and William T. Shearer

Chapter 114 Passionflower Tea Consumption and Sleep Health **227**
Alice Ngan and Russell Conduit

Chapter 115 Green Tea Consumption and Male Urinary Incontinence: A Community-based Study in Japan **229**
Andy H. Lee

Chapter 116 Greenhouse Tomato Yield and Solar Radiation **231**
Tadahisa Higashide

Chapter 117 Translocation and Distribution of Photosynthates in Tomato Plants **233**
Takashi Nishizawa and Yoshihiro Shishido

Chapter 118 Sensing, Control and Instrumentation during Tomato Growth **235**
Federico Hahn

Chapter 119 Libraries of Tomato Mutants Induced by Heavy-Ion Beam Irradiation: New Effective Method for Analyzing Functional Genomics **237**
Shunsuke Imanishi, Arisa Noguchi and Ichiro Honda

Chapter 120 Application of Molecular Methods for Tomato Breeding: The Solanum/Clavibacter Pathosystem **239**
José Pablo Lara-Ávila, June Simpson and Ángel Gabriel Alpuche-Solís

Chapter 121 Biocontrol of Fungal Diseases of Tomato: Contribution of *Pseudomonas* sp. PCI2 **241**
Nicolás A. Pastor, Susana B. Rosas, Javier A. Andrés, Marina C. Niederhauser and Marisa Rovera

Chapter 122 Begomovirus Disease: The Most Devastating Disease of Cultivated Tomato **243**
Walter N. Leke, Abwe N. M. Dione, Sali A. Ndindeng and Noé Woin

Chapter 123 Tomato Antioxidant Content as Influenced by Conventional and Organic Production Methods and Post Harvest Management **245**
Ruplal Choudhary and S. Alan Walters

Chapter 124 Sustainable Integrated Disease and Pest Management in Commercial Tomato Production in Open Fields **247**
Bombiti Nzanza, Kgabo Pofu and Phatu Mashela

Chapter 125 Trends in Foreign Trade Competitiveness for Fresh Tomato **249**
Jaime de Pablo, Miguel Giancintiv and Juan Uribe

Chapter 126 The Chemical Composition and Nutritional Value of Tomatoes **251**
M. Hernández Suárez, E. M. Rodríguez Rodríguez and C. Díaz Romero

Chapter 127 Factors Affecting the Antioxidant Content of Fresh Tomatoes and their Processed Products **253**
Norma Francenia Santos-Sánchez, Rogelio Valadez-Blanco and Raúl Salas-Coronado

Chapter 128 Lycopene in Tomato Fruits and its Content Variation during the Ripening Process after Harvest **255**
Francelina Costa, Maria de Lurdes Baeta, David Saraiva, Maria da Conceição Castilho, Manuel Teixeira Verissimo and Fernando Ramos

Chapter 129 Health Properties of Tomato Antioxidants **257**
Giulia Graziani, Antonello Santini and Alberto Ritieni

Chapter 130 The Protective Effect of Tomato Juice on the Respiratory System **259**
G. S. Maritz and M. Mutemwa

Chapter 131 Trace Elements: Effect on Tomato Plant and on Human Health After Consumption of Tomato Fruit and Tomato Fruit Food Products **261**
Charalampos Proestos, Ioannis N. Pasias, Vassiliki Papageorgiou, Konstantinos Barmperis and Nikolaos S. Thomaidis

Chapter 132 Nutraceuticals Recovery from Tomato Processing Waste and by-Products: Lycopene **263**
Antonello Santini, Giulia Graziani and Alberto Ritieni

PREFACE

This new book compiles research summaries from top professionals in the field of food sciences with a number of different focuses in this important field.

In: Food Science Research Summaries. Volume 3
Editor: Lucille Monaco Cacioppo
ISBN: 978-1-63117-864-1

Chapter 1

TECHNOLOGICAL EVALUATION OF PAN BREAD PRODUCED WITH THE ADDITION OF COMMERCIAL RESISTANT STARCH OR UNRIPE BANANA FLOUR

Caroline Joy Steel*[1]*, Camila Batista da Silva*[1]*,
Rita de Cássia Salvucci Celeste Ormenese*[1,2]*,
Maria Teresa Pedrosa Silva Clerici*[1,3]*,
***Miluska Castilho Cáceres*[1] *and Yoon Kil Chang*[1]**
[1]University of Campinas, Faculty of Food Engineering,
Department of Food Technology, Campinas, SP, Brazil
[2]Food Technology Institute, Campinas, SP, Brazil
[3]Federal University of Alfenas, Faculty of Nutrition, Alfenas, MG, Brazil

RESEARCH SUMMARY

Bread-enrichment is an excellent means of increasing the availability of functional ingredients to the general population. With the aim of making pan bread more beneficial to health, the substitution of wheat flour (WF) at levels of 0, 10, 15 e 20% by commercial resistant starch (WF, RS1, RS2, RS3) or by unripe banana flour (WF, UBF1, UBF2, UBF3) was studied. The chemical composition of WF, RS and FB, including the content of resistant starch of the two latter, was determined and the influence of the different levels of substitution on the rheological properties of WF and on the technological quality of pan bread (specific volume, moisture content, water activity and instrumental texture) was evaluated. The results were analyzed by ANOVA and the Tukey test, at a 95% significance level, and by Principle Component Analysis (PCA). RS reduced the Falling Number of WF and UBF did not affect it. Substitution of WF by increasing amounts of RS and UBF reduced dry and wet gluten contents and increased farinographic water absorption. In the alveographic analyses, they reduced dough extensibility (L) and deformation energy (W) and increased elasticity (P). Regarding pan bread quality characteristics, there was a reduction of specific volume and an increase of moisture content and hardness at all levels of substitution of WF by RS or UBF, with all breads showing an increase in hardness during the 13-day storage period, from 2.14 to 3.32 N on the first day, and reaching 7.22 to 9.30 N on the last day. According to the results obtained, the substitution of WF by 10% RS or UBF increased the content of resistant starch in pan bread and produced bread with technological characteristics similar to the standard with only WF, thus could be suggested to introduce this component in the diet.

In: Food Science Research Summaries. Volume 3
Editor: Lucille Monaco Cacioppo
ISBN: 978-1-63117-864-1

Chapter 2

IMPORTANCE OF FRENCH-TYPE BREAD IN BRAZILIAN DIET

Maria Teresa Pedrosa Silva Clerici[1], Elizabeth Harumi Nabeshima[2], Jorge Minoru Hashimoto[2], Luciano Bruno de Carvalho-Silva[1] and Yoon Kil Chang[3,*]

[1]Federal University of Alfenas – UNIFAL, MG, Brazil
[2]Food Technology Institute – ITAL, Brazil
[3]Department of Food Technology – Food Engineering Faculty,
State University of Campinas – UNICAMP, Brazil

RESEARCH SUMMARY

Bread is amongst the most consumed foods in Brazil. Data show that from 10 million tons of wheat consumed in the country, more than 50% are destined to bread production. Decades ago, bread was considered as a great source of energy, which improved quality of life of population suffering from protein-calorie malnutrition. There was great stimulation for building mills and consequently the creation of small companies, known as artisan bakeries. In these bakeries, that are small businesses with a semi-industrial room, the French-type bread, which is a popular adaptation of the French bread, is produced with a low cost and destined to immediate consumption (shelf-life around 6 hours). This kind of bread comprises almost 100% of wheat destined to bread, with only 5% destined to industrially produced breads. Those breads are most consumed during morning meals and snacks. Nutritionally, bread is still the major energy source of the diet. However, the Brazilian population profile has changed, since the country's greatest problems in public health are: obesity, diabetes, hypertension and heart diseases. Therefore, French-type bread, which has high glycemic index and high content of sodium, has been contraindicated. However, the habit of consuming it is already well established in Brazilian population. With the new concepts of functional foods and health sustainability, i.e., food must contribute to the maintenance of subjects'

[*] E-mail address: mariateresapedrosa@gmail.com.

well-being, the universities and bakery industries are making all efforts to improve the nutritional quality of artisan and white bread. This chapter will provide for economic studies and studies on the technological, sensory and nutritional characteristics of French-type bread.

In: Food Science Research Summaries. Volume 3
Editor: Lucille Monaco Cacioppo
ISBN: 978-1-63117-864-1

Chapter 3

EFFECT OF ADDING ASCORBIC ACID ON THE QUALITY OF SLICED LOAF WITH PARTIAL SUBSTITUTION OF AMARANTH FLOUR (AMARANTHUS CAUDATUS) AND WHEAT BRAN (TRITICUM AESTIVUM)

Luz Maria Paucar-Menacho**, *Nilthon Santa María-Podestá and Nathaly Aguedita Cerna-Lucio
Departamento de Ingeniería Agroindustrial, Universidad Nacional del Santa, Perú

RESEARCH SUMMARY

The current consumer trend is consuming foods with health benefits and easy preparation. Functional sliced bread with addition of amaranth flour and wheat bran, in order to study the influence of adding ascorbic acid to the finished product was elaborated in this research. The formulations were made using a factorial experimental design 2^3 with 17 experiments (8 factorial experiments, 6 axial experiments and 3 replicates at center point), considering the levels of amaranth flour (*Amaranthus caudatus*), wheat bran and ascorbic acid as independent variables. The effects of these variables were evaluated on the physical-chemical characteristics (specific volume, crust and crumb instrumental color of bread) and sensory properties performed with 30 panelists. The results were analyzed by Response Surface Methodology (RSM), indicating that ascorbic acid improves the quality of bread in their characteristic of specific volume, likewise, the addition of wheat bran fiber and cereals such as amaranth flour influenced the quality demonstrated in the buying intention for formulations 3 (5% of Amaranth flour, 8.5% of wheat bran and 36 ppm of ascorbic acid) and 5 (5% of amaranth flour, 3.5% of wheat bran and 84 ppm of ascorbic acid). Wheat bran, amaranth flour and ascorbic acid had significant influence on the volume of sliced breads. The formulations 5 (5% of amaranth flour, 3.5% of wheat bran and 84 ppm of ascorbic acid)

* E-mail address: luzpaucar@uns.edu.pe.

and 9 (2% of amaranth flour, 6% of wheat bran and 60 ppm of ascorbic acid) showed the highest specific volumes; 5.02 mL/g and 5.04 mL/g, respectively, higher than the specific volume presented by the standard bread with 4.49 mL/g, due to the ascorbic acid effect. Wheat bran affected negatively the instrumental color, specific volume, and the crumb color and texture in the sensory analysis of sliced breads. The chemical score of sliced breads did not show amino acid deficit, except the amino acid lysine; but were more than 70% the amount recommended by FAO in all formulations.

In: Food Science Research Summaries. Volume 3
Editor: Lucille Monaco Cacioppo
ISBN: 978-1-63117-864-1

Chapter 4

MAIZE BREAD: HEALTHY AND SAFE

Malvar Rosa Ana, Butrón Ana and Revilla Pedro*
Misión Biológica de Galicia (CSIC), Pontevedra, Spain

RESEARCH SUMMARY

Bread is one of the bases of the Mediterranean diet, but as the society becomes wealthier, bread consumption decreases dramatically. Wheat and barley bread was part of the ancient Mediterranean civilizations, and maize was enthusiastically incorporated into the traditional European cuisine as soon as the Spaniards met this high yielding cereal in Central America. Indeed, maize was a solution to the famine crisis that devastated Europe and became one main provider of carbohydrates in the humid Spain, northern Portugal and throughout the Old World soon after the discovery of America. Whole grain is used for manufacturing bread and several derived recipes either with maize alone or in combination with other cereals, such as wheat, rye or rice, depending on the production of each particular region. Originally, maize used for bread came from tropical areas of Central America, where it had been selected by the natives for human consumption. Those first varieties were mainly white flint corn with a high content of proteins and fats and were adapted to temperate conditions. Besides carbohydrates, maize is a convenient source of minerals, vitamins and antioxidant compounds that can be worthwhile to exploit, given the high yield of maize compared to other cereals. Maize proteins are low in lysine and tryptophan, and the Native Americans used to combine maize with beans in order to get an equilibrated protein intake. There is wide variability within the maize germplasm for most of these quality parameters. From the perspective of health, two singular questions affect maize for human consumption: first, maize lacks gluten and, therefore, constitutes a key nutrient for people suffering from celiac disease; and second, we cannot neglect the problem of micotoxins, which are toxic substances produced by fungi. These toxins can reach levels that pose a threat for human health depending on growing and storing conditions. Therefore, this review focuses on three main issues of maize bread: germplasm and breeding, quality and food safety.

* E-mail address: previlla@mbg.csic.es.

In: Food Science Research Summaries. Volume 3
Editor: Lucille Monaco Cacioppo

ISBN: 978-1-63117-864-1

Chapter 5

NUTRITIONAL QUALITY OF PROTEINS IN BREADS

***Luciano Bruno de Carvalho-Silva*[1]**
***and Maria Teresa Pedrosa Silva Clerici*[2]**

[1]Federal University of Alfenas- MG (UNIFAL-MG); Center for Food Security Studies (NEPA), State University of Campinas (UNICAMP), Brazil
[2]Federal University of Alfenas – UNIFAL, MG, Brazil

RESEARCH SUMMARY

Bread may be an optimal source for protein enrichment, as wheat proteins have quantitative deficiencies in some essential amino acids. Works focusing on the use of new protein sources in breads present a major concern about technological quality, however, few studies perform a full nutritional evaluation of the final product. In this chapter the main raw materials used to enhance bread protein value will be showed, where the amino acid profile will be discussed and the main works focusing on the nutritional quality of protein of bread will be presented.

In: Food Science Research Summaries. Volume 3
Editor: Lucille Monaco Cacioppo
ISBN: 978-1-63117-864-1

Chapter 6

FIBER-ENRICHED BREAD AS AN AID IN THE NUTRITIONAL MANAGEMENT OF DISLIPEMIC PATIENTS

***Ismael Darío Bianco*[1,2,3,*], *Dante Miguel Beltramo*[1,2,4] and *Gustavo Tomás Diaz*[5]**

[1]Centro de Excelencia en Productos y Procesos de Córdoba (CEPROCOR)
[2]Consejo Nacional de Investigaciones Científicas y Técnicas (CONICET)
[3]Facultad de Ciencias Químicas, Universidad Católica de Córdoba, Argentina
[4]Departamento de Ciencias Exactas, Físicas y Naturales, Universidad Nacional de La Rioja, Argentina
[5]Centro de Salud San Ricardo Pampuri, Villa Carlos Paz, Argentina

RESEARCH SUMMARY

In the XXI century the main epidemics are not infectious diseases such as cholera, bubonic plague or AIDS, but chronic diseases such as obesity, diabetes, atherosclerosis and cancer. What these conditions have in common is that all have been linked to nutrition and lifestyle. In this context, it has been shown that nutritional fiber can modify the glycemic index inducing a slow and steady absorption of food and increasing the secretion of incretins by the digestive system. Furthermore, soluble fiber is important to decrease the absorption of bile acids and cholesterol contributing in the prevention of dislipemy and colon cancer. Nutritional fiber can also decrease the caloric density of food aiding to weight control.

Bread is a food with special characteristics because it is widely accepted and can be manipulated to change its composition so as to be complete, accessible and enriched with nutritional elements. It has been shown that patients receiving chitosan-containing bread decreased their mean levels of LDL cholesterol and significantly increased their mean levels of HDL cholesterol. Recent studies in bread making include the addition of chia flour as a

* E-mail address: ibianco@ceprocor.uncor.edu; ibianco@hotmail.com

gluten-free source of omega three fatty acids, soluble fiber and protein. Hydrocolloids have also been employed as modifiers of dough volume and rheology.

In: Food Science Research Summaries. Volume 3
Editor: Lucille Monaco Cacioppo
ISBN: 978-1-63117-864-1

Chapter 7

EFFECT OF DIFFERENT PINEAPPLE JUICE (*ANANAS COMOSUS* L.) PREPARATIONS ON THE MICROSTRUCTURE, STALING AND TEXTURAL PROPERTIES OF WHEAT BREAD

***Ola Lasekan*[1,*]*, Chinma Chiemela*[2]*, Beatrice Ossai*[2]**
***and Noranizan Mohd Adzahan*[1]**
[1]University Putra Malaysia, Faculty of Food Science and Technology, Selangor, Malaysia
[2]Federal University of Technology, Department of Food Nutrition, Minna, Niger State, Nigeria

RESEARCH SUMMARY

The effect of adding different pineapple juice preparations on the microstructure, staling and textural properties of control wheat bread (WB), and concentrated pineapple juice (PJC) and pineapple-juice breads (PJ) respectively were investigated. The loaves quality, crust and crumb color, staling profile (crumb firmness, enthalpy) during storage were monitored. The micro structural differences of crumbs were also examined using microtome and Scanning Electron Microscopy (SEM). Results revealed that the specific volume of fresh loaves baked with PJC) was not significantly ($P < 0.05$) different from that of (WB). Likewise, firmness of crumbs, number of crumb cells, area air cell, bake loss, volume yield, crust and crumb color of the freshly baked PJC were not significantly ($P < 0.05$) different from the WB. However, during storage the retrogradation of starch in PJC and PJ was significantly ($P < 0.05$) retarded as compared with the WB. While there was no significant ($P < 0.05$) textural differences between the crumbs of freshly produced PJC and the crumb of the WB, staled breads revealed significant textural changes in their crumbs. Microtome and SEM revealed the presence of many starch granules on the surface of pore walls in the WB, whereas the PJC contained

* E-mail: lasekan61156@yahoo.com.

starch gels cemented between starch granules. Gluten nets were found to be uniform and more oriented in the WB than in the PJC. However, same gluten nets were dense and non-uniform in the PJ.

In: Food Science Research Summaries. Volume 3
Editor: Lucille Monaco Cacioppo
ISBN: 978-1-63117-864-1

Chapter 8

NUTRITIONAL AND TECHNOLOGICAL CHARACTERISTICS OF GLUTEN-FREE AND LOW GLYCEMIC INDEX BAKING PRODUCTS

A. M. Calderón de la Barca[1,*], M. I. Ortega-Vélez[1] and F. Cabrera-Chávez[2]

[1]Coordinación de Nutrición. Centro de Investigación en Alimentación y Desarrollo A.C. Hermosillo, Sonora, México

[2]Dipartimento di Scienze Molecolari Agroalimentari. Università degli Studi di Milano. Milano, Italia

RESEARCH SUMMARY

Celiac disease (CD) and diabetic patients require life-long special diets. Accordingly, CD is an enteropathy triggered by wheat gluten, thus patients must follow a strict gluten-free diet avoiding gluten from wheat, rye and barley. On the other hand, diabetic patients require low glycemic index foods for a low blood glucose response after consumption. In order to face their need, several breads have been formulated and commercialized for gluten-free and low glycemic index diets. New developing of these special breads has centered on their technological and sensorial properties and few special bread formulations have been advocated to meet nutritional requirements for different age groups who suffer from CD or diabetes. This becomes a very important issue because some nutrients and energy are mostly taken from wheat-based foodstuffs, principally bread, in the current western diet. Another concern is that industrialized special breads are more expensive and less available at the market than the current wheat-based products. Therefore, breads for CD and diabetic patients are a technological and nutritional challenge, since they should offer optimum characteristics of functional quality and high nutritional value at acceptable costs. In this chapter, trends in special bread formulations focusing on their technological properties and some nutritional aspects are discussed.

---

* E-mail address: amc@ciad.mx.

In: Food Science Research Summaries. Volume 3
Editor: Lucille Monaco Cacioppo

ISBN: 978-1-63117-864-1

Chapter 9

EXPERIMENTAL DESIGN FOR THE EVALUATION OF BREAD PROTEIN QUALITY

***Luciano Bruno de Carvalho-Silva*[1,*]**
***and Maria Teresa Pedrosa Silva Clerici*[2*]**
[1]Federal University of Alfenas- MG (UNIFAL-MG); Center for Food Security Studies (NEPA), State University of Campinas (UNICAMP), Brazil
[2]Federal University of Alfenas – UNIFAL-MG, Brazil

RESEARCH SUMMARY

The evaluation of protein quality is a key factor in determining nutritional and commercial value of food products. Techniques based mainly on the protein efficacy in stimulating animal growth have been developed and used for this purpose. Many works in the area of breadmaking have tried to complement the protein in wheat flour, which presents deficiencies in essential amino acids, with different protein sources, however, very few studies focus on the nutritional value. This chapter presents a proposal of nutritional studies which may be used to evaluate the nutritional quality of proteins, presenting some examples of works which have been applied in bread evaluation. Furthermore, reports the classic literature on the main methods of protein quality evaluation.

* E-mail address: lucianobster@gmail.com.

In: Food Science Research Summaries. Volume 3
Editor: Lucille Monaco Cacioppo
ISBN: 978-1-63117-864-1

Chapter 10

PACKAGING AND REQUIRED PROTECTION FOR THE PRESERVATION OF BREAD FUNCTIONAL COMPOUNDS

Eduardo Henrique Miranda Walter*
Embrapa Food Technology, Rio de Janeiro, Brazil

RESEARCH SUMMARY

Requirements to stabilize or enhance bread functional value are fast becoming major market demands. Packaging plays an essential role in the protection of the wholesomeness, sensory and nutritional qualities of bread, especially in products which require an extended shelf life and are formulated with sensitive functional ingredients. In the face of new products development, this chapter presents packaging technologies as part of an integrated strategy to preserve bread functional compounds.

* E-mail address: eduardo.walter@ctaa.embrapa.br.

In: Food Science Research Summaries. Volume 3
Editor: Lucille Monaco Cacioppo
ISBN: 978-1-63117-864-1

Chapter 11

GASTROINTESTINAL HEALTH PROMOTION VIA BREAD CONSUMPTION: ELUCIDATION OF PREVENTIVE EFFECTS IN THE ETIOLOGY OF PATHOLOGICAL EVENTS ALONG THE INTESTINE

Beate Hiller** Wiebke Schlörmann, Jürgen Hollmann, Michael Glei and Meinolf Lindhauer
Department of Safety and Quality of Cereals, Max Rubner-Institute, Detmold, Germany

RESEARCH SUMMARY

Dramatically increasing incidences of inflammatory and malignant diseases of the digestive tract have urged the identification of dietary factors/patterns of chemopreventive potential in the etiology of these mainly Western lifestyle associated diseases.

Investigations demonstrated that bread does not only constitute major staple food product in our everyday day diet, but does fundamentally contribute to gastrointestinal epithelial health maintenance by

- serving as naturally rich source of complex carbohydrates, dietary fiber, fiber-bound mono-/diphenolics, antioxidants as well as phytic acid (in amounts reflecting cereal source and bread composition) as well as processing-related antioxidants such as pronyllysine and hydroxymethyl-furfural formed during the baking process;
- affecting gastrointestinal metabolite formation from exogenous/endogenous compounds, stimulating the release of chemopreventive agents such as acetate, propionate, butyrate, lactate, free phenolics well as phenolic acid metabolites, and limiting the production of toxic amino and bile acid metabolites as well as microbial enzyme proteins;

* E-mail address: beate.hiller.sci@gmx.de.

- acting as prebiotic substrates for beneficial intestinal microbiota;
- exerting various chemopreventive effects along the intestine.

This chapter summarizes recent research data as well as reviewed literature findings concerning gastrointestinal health promotion via bread consumption, outlining preventive effects in the etiology of early- and progressed-stage pathological events along the intestine.

In: Food Science Research Summaries. Volume 3
Editor: Lucille Monaco Cacioppo
ISBN: 978-1-63117-864-1

Chapter 12

WHAT IS COFFEE?

Bernard A. Goodman[*]
State Key Laboratory for Conservation and Utilization of Subtropical Agro-Bioresources,
Guangxi University, Nanning, Guangxi, China

RESEARCH SUMMARY

This chapter provides an introduction to the diversity of the various types of coffee and the variability in the contents of their principal chemical components. Attention is drawn to factors which affect the chemical composition of the beverage and its aroma, including plant genetics, production conditions of the green beans, roasting, and brewing. Such recognition of the range in composition of the various types of coffee represents essential background information on which to evaluate the sometimes conflicting results from epidemiological investigations that form the major part of this book.

[*] E-mail: bernard_a_goodman@yahoo.com.

In: Food Science Research Summaries. Volume 3
Editor: Lucille Monaco Cacioppo
ISBN: 978-1-63117-864-1

Chapter 13

ANTIOXIDANT AND FREE RADICAL PROCESSES IN COFFEE

Bernard A. Goodman*
State Key Laboratory for Conservation and Utilization of Subtropical Agro-Bioresources, Guangxi University, Nanning, Guangxi, China

RESEARCH SUMMARY

Free radical processes and associated antioxidant reactions are of fundamental importance in biology, and have been implicated in the etiology of a wide range of disease processes. Since many antioxidant molecules are derived from the foods we eat, there has in recent years been an increasing trend to define food quality in terms of its antioxidant capacity.

The concepts of antioxidant capacity and free radical scavenging are discussed in this chapter along with chemical processes in coffee that specifically involve free radicals and antioxidant molecules. Consideration is given to reactions that occur during roasting and the storage of roasted coffee beans, as well as to processes that occur in the brewed beverage, and the implications of such reactions are discussed in terms of coffee quality criteria.

* E-mail: bernard_a_goodman@yahoo.com.

In: Food Science Research Summaries. Volume 3
Editor: Lucille Monaco Cacioppo
ISBN: 978-1-63117-864-1

Chapter 14

Properties and Analysis of Caffeine and Chlorogenic Acids in Coffee Beans

Abebe Belay
Adama Science and Technology University, School of Natural Sciences, Physics Department, Adama, Ethiopia.

Research Summary

The chemical characteristics, composition, and the physiological and psychological effects of caffeine and chlorogenic acids in biological systems are presented, along with their relationships with coffee quality. In addition, various chemical and physical methods that can be used for the routine analysis of the contents of these compounds in coffee beans are reviewed.

In: Food Science Research Summaries. Volume 3
Editor: Lucille Monaco Cacioppo
ISBN: 978-1-63117-864-1

Chapter 15

Occurrence and Identification of Chlorogenic Acids in Green and Roasted Coffee Beans

Nikolai Kuhnert and Rakesh Jaiswal
Jacobs University, Bremen, Germany

Research Summary

Second only to water, coffee constitutes in many countries, such as the USA and Germany, the most popular beverage. On a global scale coffee follows water and black tea as the third most heavily-consumed beverage, with an annual production of around 8 Mt and an average daily consumption of 2.3 billion cups a day (Higdon and Frei 2006).

Coffee is obtained from the seeds of the cherries of *Caffea Arabica or Caffea canephora*, otherwise referred to as green coffee beans. To obtain the popular beverage, green coffee beans are roasted at temperatures ranging between 180 °C and 250 °C with roasting times ranging from 5 to 30 minutes, producing so called light to dark roasts (Bekedam *et al.* 2008).

The green coffee bean contains next to its primary metabolites, such as carbohydrates, proteins and lipids, a wide variety of secondary metabolites. Caffeine is certainly the best known of these secondary metabolites, but the class of compounds termed chlorogenic acids (CGAs) dominates the secondary metabolite profiles of green coffee beans, both in terms of number of compounds present and their weight.

Similarly CGAs play an important role in coffee roasting, reacting in Maillard type reactions to form simple products, and also the polymeric melanoidins, which are responsible for the sensory properties of a cup of coffee.

Finally, CGAs are responsible for a wide variety of beneficial health effects derived from consumption of the coffee beverage, as shown by a series of epidemiological and clinical studies (Higdon and Frei 2006).

This chapter discusses the occurrence of chlorogenic acids in green and roasted coffee, and discusses in depth the analytical methods, mainly based on tandem mass spectrometry, that are used to identify individual compounds.

In: Food Science Research Summaries. Volume 3
Editor: Lucille Monaco Cacioppo
ISBN: 978-1-63117-864-1

Chapter 16

COFFEE ROASTING: ACCURATE CONTROL FOR INCREASED BIOACTIVITY

Susana Casal*
REQUIMTE, Laboratório de Bromatologia e Hidrologia,
Faculdade Farmácia, Universidade do Porto, Portugal

RESEARCH SUMMARY

To reveal its flavour, coffee has to be roasted. Green coffee beans are almost devoid of aroma, and it is only after roasting that coffee's characteristic appearance, taste and aroma are developed. Considering the importance of roasting, understanding the resulting physical and chemical transformations, are of utmost importance to coffee quality.

During the roasting process, the most critical phase in the coffee production chain, some of the chemical components of green beans are loss, whilst others react and produce new compounds, thus making roasted coffee one of the most complex foods from the chemical point of view.

The composition of the final roasted coffee is defined by a multiplicity of factors: the roasting parameters on one side and the green beans characteristics on the other. Furthermore, the agronomic practices used in the production of coffee berries, post-harvest processing during their conversion into commercial green coffee, and even its handling during transportation and storage, determine the chemical composition of the raw materials. Then the equipment used for roasting, and the temperature/time regimen applied to the beans establish the final characteristics of the product.

This chapter focuses on recent developments in the control of the roasting parameters to enhance the contents of various bioactive components in roasted coffee beans, and hence its potential health effects.

* E-mail: sucasal@ff.up.pt.

In: Food Science Research Summaries. Volume 3
Editor: Lucille Monaco Cacioppo
ISBN: 978-1-63117-864-1

Chapter 17

CHARACTERIZATION OF ORGANIC COFFEE USING PHOTOTHERMAL TECHNIQUES

F. Gordillo-Delgado[a,b,*] ***and E. Marín***[b]

[a]Laboratorio de optoelectrónica, Universidad del Quindío,
Apdo. 2639 Armenia, Colombia
[b]Centro de Investigación en Ciencia Aplicada y Tecnología
Avanzada del Instituto Politécnico Nacional, Unidad Legaria,
Legaria 694. Col. Irrigación, México D.F, México

RESEARCH SUMMARY

This chapter describes various photothermal techniques and reviews some of their applications for organic coffee characterization. These techniques measure changes in the thermodynamic state of a sample resulting from periodically modulated heating. Photoacoustic spectroscopy avoids the light dispersion problems that can interfere with conventional spectrophotometric methods, and the present paper focuses on the characterization of grains, powders and infusions of roasted and soluble coffee.

Furthermore, photoacoustic is a non-invasive and inexpensive technique that allows investigations of in vivo and in situ photosynthetic activity in plants. Examples of the monitoring of photosynthetic activity of coffee plants are presented, along with the use to photoacoustic technique to show the effects of environmentally friendly bacterial inoculants.

In addition, knowledge of thermal parameters is particularly important for improving the industrial roasting of coffee as well as for monitoring the degradation of the final product when, for example, it is exposed to cyclic periods of heating that can affect its quality. Finally, we also show how the photosynthetic activity of coffee plants, the thermal parameters of coffee beans, and the presence of pigments and contents of certain chemical components in ground and roasted coffee differ according to plant growth conditions.

* E-mail: fgordillo@uniquindio.edu.co.

In: Food Science Research Summaries. Volume 3
Editor: Lucille Monaco Cacioppo
ISBN: 978-1-63117-864-1

Chapter 18

COFFEE CONSUMPTION AND THE HEALTH OF THE GASTROINTESTINAL AND ENDOCRINE SYSTEMS

Bernard A. Goodman*
State Key Laboratory for Conservation and Utilization of Subtropical Agro-Bioresources, Guangxi University, Nanning, Guangxi, China

RESEARCH SUMMARY

This chapter describes briefly the digestive process, and the epidemiological evidence for the effects of coffee on various organs that are involved in the gastrointestinal and endocrine systems. These include separate sections dealing with the possible impact of coffee consumption on processes in the upper and lower digestive tracts, including its possible contribution to the breakdown of complex food structures; this is supplemented by information relating to the health of the pancreas, liver and kidneys. Speculation as to the chemical basis for some of these observed effects is discussed, and some indication given of areas in which more work is needed for a better understanding of the roles of coffee in the health of organs in these systems.

* E-mail: bernard_a_goodman@yahoo.com.

In: Food Science Research Summaries. Volume 3
Editor: Lucille Monaco Cacioppo
ISBN: 978-1-63117-864-1

Chapter 19

EFFECTS OF COFFEE ON INFLAMMATORY PROCESSES

Bernard A. Goodman*

State Key Laboratory for Conservation and Utilization of Subtropical Agro-Bioresources, Guangxi University, Nanning, Guangxi, China

RESEARCH SUMMARY

Since inflammation is directly associated with many disease states, both infectious and non-infectious, this chapter provides a brief summary of the evidence that links coffee consumption with inflammatory processes. Inflammation is a natural stage in the normal healing process, but it can lead to persistent damage if it gets out of control. Current evidence suggests that coffee is able to provide some protection against inflammatory damage from non-infectious diseases in some persons, but not from infectious diseases.

* E-mail: bernard_a_goodman@yahoo.com.

In: Food Science Research Summaries. Volume 3
Editor: Lucille Monaco Cacioppo

ISBN: 978-1-63117-864-1

Chapter 20

EPIDEMIOLOGICAL EVIDENCE FOR THE ASSOCIATION OF COFFEE CONSUMPTION WITH CANCER

Bernard A. Goodman*

State Key Laboratory for Conservation and Utilization of Subtropical Agro-Bioresources, Guangxi University, Nanning, Guangxi, China

RESEARCH SUMMARY

This chapter briefly reviews the association of coffee consumption and cancers of various organs in the body, with an emphasis on the links identified from human epidemiological studies. In *in vitro* assays, coffee and several of its components exhibit mutagenic activity, but there is little positive association of coffee consumption with increased cancer risk. On the contrary, coffee consumption appears to provide some protection against the development of tumors in the brain and liver, as well as against colorectal, endometrial and prostate cancers, and leukemia. In addition coffee consumption has little effect on bladder, breast, gastric, oral, ovarian, and renal cancers, and there is conflicting evidence of its association with the risk of lung and pancreatic cancers. The epidemiological evidence for these observations is presented along with a discussion of the chemical components of coffee that might be involved in these effects.

* E-mail: bernard_a_goodman@yahoo.com.

In: Food Science Research Summaries. Volume 3
Editor: Lucille Monaco Cacioppo

ISBN: 978-1-63117-864-1

Chapter 21

COFFEE CONSUMPTION AND OXIDATIVE DNA DAMAGE

Takeshi Hirano[1,*], Haruki Morii[2] and Kazuyoshi Tamae[3]

[1]Department of Life and Environment Engineering, Faculty of Environmental Engineering, University of Kitakyushu, Wakamatsu-ku, Kitakyushu, Fukuoka, Japan
[2]Graduate School of Environmental Engineering, University of Kitakyushu, Wakamatsu-ku, Kitakyushu, Fukuoka, Japan
[3]Division of Teacher Training, Faculty of Education and Culture, University of Miyazaki, Miyazaki, Japan

RESEARCH SUMMARY

Many studies have investigated the role of dietary factors on human health, including the effects of coffee consumption. Although coffee has been a popular beverage for a long time, its exact health effects are unknown and remain controversial. Some studies have reported that coffee consumption lowers the risk of some types of cancers, whereas others provide negative or inverse conclusions for the effects of coffee consumption on human health.

To analyze the effects of instant coffee consumption on cancer risk, we examined the association between instant coffee consumption and the generation in mice of oxidative DNA damage. 7,8-dihydro-8-oxoguanine (8-oxoguanine, 8-oxo-Gua). 8-Oxo-Gua is a major form of oxidative DNA damage, and is believed to play a key role in carcinogenesis, but our study showed no significant differences between water drinking and coffee drinking, in terms of 8-oxo-Gua generation in the liver. These results suggest that instant coffee consumption has little, if any, effect on the risk of liver cancer due to oxidative stresses.

* Corresponding author. Tel: +81-93-695-3206, Fax: +81-93-695-3299, E-mail address: t-hirano@kitakyu-u.ac.jp.

In: Food Science Research Summaries. Volume 3
Editor: Lucille Monaco Cacioppo ISBN: 978-1-63117-864-1

Chapter 22

COFFEE, CAFFEINE AND THE CARDIOVASCULAR SYSTEM

Bernard A. Goodman*
State Key Laboratory for Conservation and Utilization of Subtropical Agro-Bioresources, Guangxi University, Nanning, Guangxi, China

RESEARCH SUMMARY

This chapter briefly reviews the impact of caffeine and coffee consumption on the health of the cardiovascular system. One of the detrimental effects of caffeine consumption is a raising of the blood pressure (hypertension), but coffee contains components which to a large extent neutralise this effect of caffeine, and prospective studies have not established any association of coffee consumption with deaths from coronary heart disease. Furthermore, any links between coffee consumption and serum cholesterol is strongly dependent on the type of coffee and the brewing method, since instant and paper-filtered coffees have low contents of coffee lipids, whereas this component is retained in other forms of coffee. In addition, coffee consumption seems to provide some protection against the risk of stroke. Thus in contrast to the often perceived detrimental effects of coffee on the cardiovascular system, the current evidence is that for the majority of people its net effects are largely neutral and possibly beneficial.

* E-mail: bernard_a_goodman@yahoo.com.

In: Food Science Research Summaries. Volume 3
Editor: Lucille Monaco Cacioppo
ISBN: 978-1-63117-864-1

Chapter 23

COFFEE CONSUMPTION AND TYPE 2 DIABETES MELLITUS

Juliane Costa Silva Zemdegs*[1,2]*, Cristiane de Oliveira*[3]*,
***Anita Sachs*[2] *and João Felipe Mota*[1,4]**

[1]Department of Physiology, Division of Nutrition Physiology,
Federal University of Sao Paulo, Brazil
[2]Department of Preventive Medicine, Division of Nutrition,
Federal University of Sao Paulo, Brazil
[3]Systemic Inflammation Laboratory, Trauma Research,
St. Joseph's Hospital and Medical Center, Phoenix, Arizona
[4]Nutrition Faculty, Federal University of Goiás, Brazil

RESEARCH SUMMARY

The prevalence of type 2 diabetes mellitus has increased in recent years in both developed and developing countries, and several risk factors and co-morbities have been described. Among the physiopathological mechanisms for the disease, oxidative stress and inflammation contribute to both its development and complications. However, the bioactive components of coffee, non-phenolic (caffeine, trigonelline, nicotinic acid) and phenolic compounds (chlorogenic acid and caffeic acid) have been associated with effects on glucose metabolism, and a possible protective effect of coffee on the risk of type 2 diabetes mellitus has emerged. This chapter describes the epidemiological data relating coffee consumption and the risk for type 2 diabetes mellitus, and also presents and discusses the mechanisms that have been proposed for the beneficial effects of coffee and its bioactive components on glucose metabolism.

In: Food Science Research Summaries. Volume 3
Editor: Lucille Monaco Cacioppo
ISBN: 978-1-63117-864-1

Chapter 24

NEUROLOGICAL EFFECTS OF COFFEE CONSUMPTION

Bernard A. Goodman*
State Key Laboratory for Conservation and Utilization of Subtropical Agro-Bioresources, Guangxi University, Nanning, Guangxi, China

RESEARCH SUMMARY

Caffeine is a widely used psychoactive substance, and for many people coffee is the main source of caffeine in their diet. Several beneficial psychological effects have been associated with caffeine consumption, including mental stimulation, improvement in moods, and analgesic properties. In addition, coffee consumption is associated with decreased risks of neurodegenerative diseases, such Parkinson's disease, Alzheimer's disease, and dementia. On the other hand caffeine intake is also associated with sleep disorders and anxiety symptoms. The current status of research on these various actions of coffee is presented in this chapter.

* E-mail: bernard_a_goodman@yahoo.com.

In: Food Science Research Summaries. Volume 3
Editor: Lucille Monaco Cacioppo
ISBN: 978-1-63117-864-1

Chapter 25

EFFECTS OF COFFEE CONSUMPTION DURING PREGNANCY AND IN YOUNG CHILDREN

Bernard A. Goodman*

State Key Laboratory for Conservation and Utilization of Subtropical Agro-Bioresources, Guangxi University, Nanning, Guangxi, China

RESEARCH SUMMARY

This chapter discusses the various claims for links between caffeine/coffee consumption during pregnancy and problems relating to the fetal development, new-born and young children. Items studied include the risk of miscarriage, fetal development and birth weight, stillbirth and birth defects, breastfeeding, and sudden infant death syndrome (SIDS). Most are still controversial; some studies indicate positive links with caffeine/coffee consumption, whilst others show none after allowing for confounding factors, such as cigarette smoking, alcohol consumption, etc. However, there does appear to be no benefit to be derived from coffee consumption during pregnancy and whilst breastfeeding. Also, since the rate of caffeine metabolism slows down during pregnancy, it is probably prudent to decrease consumption of caffeinated beverages.

* E-mail: bernard_a_goodman@yahoo.com.

In: Food Science Research Summaries. Volume 3
Editor: Lucille Monaco Cacioppo
ISBN: 978-1-63117-864-1

Chapter 26

EFFECTS OF CAFFEINE CONSUMPTION ON EXERCISE

Bernard A. Goodman*
State Key Laboratory for Conservation and Utilization of Subtropical Agro-Bioresources, Guangxi University, Nanning, Guangxi, China

RESEARCH SUMMARY

Many studies have shown that consumption of caffeine enhances the ability to perform endurance exercise, and the evidence to support this is reviewed in the first part of this chapter. Caffeine acts as a general antagonist of adenosine receptors in the brain, producing an analgesic effect which decreases the impact of exercise-induced pain. This then allows athletes to train/perform at higher speed or for a longer time in aerobic exercise. Caffeine consumption also results in an increase in blood pressure and a decreased heart rate, and may also have a physiological effect by enhancing the muscular contractile force. The impact of caffeine on short-term high impact performance is less clear-cut, and may depend on the type of exercise as well as the degree of training of the subject. Little work has been performed on the effects of caffeine consumption on exercise in children, but the evidence suggests that the effects on blood pressure and heart rate are similar to those observed with adults.

* E-mail: bernard_a_goodman@yahoo.com.

In: Food Science Research Summaries. Volume 3
Editor: Lucille Monaco Cacioppo

ISBN: 978-1-63117-864-1

Chapter 27

OTHER HEALTH-RELATED ASPECTS OF COFFEE CONSUMPTION

Bernard A. Goodman*

State Key Laboratory for Conservation and Utilization of Subtropical Agro-Bioresources, Guangxi University, Nanning, Guangxi, China

RESEARCH SUMMARY

This chapter briefly reviews health-related issues that have been linked with coffee consumption. Short sections are presented dealing with the effects of coffee consumption on asthma, rheumatoid arthritis and gout, gallbladder disease, fibrocystic breast disease, dental health, laxative/diuretic effects, bone health, and sexual behavior. For most of these health issues, the relationships are not clear-cut, but it appears that, at least for some people, coffee provides some protection against allergic asthma, gout, gallstone disease, and oral health, whereas it is ineffective against other forms of rheumatoid arthritis, and it may contribute to the development of fibrocystic breast disease. Finally, coffee can have a laxative/diuretic effect, and (in rats) it has been shown to decrease erectile dysfunction, and to increase sexual activity in females.

* E-mail: bernard_a_goodman@yahoo.com.

In: Food Science Research Summaries. Volume 3
Editor: Lucille Monaco Cacioppo
ISBN: 978-1-63117-864-1

Chapter 28

EFFECT OF IONIZING RADIATION ON POLYMERIC FOOD PACKAGING MATERIALS

A. Badeka and M. G. Kontominas*
Laboratory of Food Chemistry, Department of Chemistry, University of Ioannina, Ioannina, Greece

RESEARCH SUMMARY

Ionizing radiation is currently being used in over 40 countries for food preservation purposes and in many more for the sterilization of packaging materials, pharmaceutical and medical products. Foodstuffs to be irradiated are usually prepackaged to avoid recontamination. Thus, packaging materials are affected in several ways by ionizing radiation. Changes manifested as polymer scission or crosslinking affect key properties of polymeric packaging materials such as production of radiolysis products, migration, barrier and mechanical properties which in turn may affect the sensory properties and safety of the packaged food product.

The present article reviews the literature on the subject with emphasis on the work carried out in the laboratory of the authors during the past 20 years.

* E-mails: mkontomi@cc.uoi.gr; abadeka@cc.uoi.gr.

In: Food Science Research Summaries. Volume 3
Editor: Lucille Monaco Cacioppo

ISBN: 978-1-63117-864-1

Chapter 29

PACKAGING OPTIMIZATION FOR MINIMALLY PROCESSED FRUIT AND VEGETABLES

***Amalia Conte*[1], *Marcella Mastromatteo*[2] and *Matteo Alessandro Del Nobile*[3]**
University of Foggia, Agricultural Faculty,
Department of Food Science, Foggia, Italy

RESEARCH SUMMARY

Minimally processed crops consist in washed, cut and packaged fruit and vegetables. Consumer preferences are increasing towards these ready to use products, thus promoting a great effort of research in this area. The value of fresh-cut products lies on the primary characteristics of freshness and convenience. However, operations such as peeling, cutting, shredding or slicing greatly increase tissue damage of fresh-cut commodities. Various food preservation methods have been developed over the years. Today, the interest towards a green consumerism is increasing around the world. Therefore, as valid substitute of chemicals, natural compounds with antimicrobial and antioxidant properties are highly proposed for dipping or coating fresh-cut crops. Modified atmosphere packaging (MAP) is another preservation method to maintain the natural quality of minimally processed food. MAP for fruit and vegetables can be created either passively, by the product respiration, or with intent, by substituting air with a proper gas mixture in the package. It is well recognized that success of MAP greatly depends on the accuracy of predictive respiration rate models. The combination of natural antimicrobials to MAP conditions in a sealed packaging system represents a strategic solution to prolong the shelf life of several fresh-cut products. The most relevant aspect to be taken into account for maintaining the quality of horticultural commodities is the choice of the appropriate packaging system. Different headspace conditions can be achieved in the package depending on the interactions between respiratory

[1] Tel +39 0881 589 240, a.conte@unifg.it.
[2] Tel +39 0881 338 575, ma.mastromatteo@unifg.it.
[3] Tel +39 0881 589 242, ma.delnobile@unifg.it.

activity of the packaged produce and gas transfer through the polymeric film. Therefore, the proper choice of film mass barrier properties is a key factor in order to obtain optimum modification of the atmosphere and to avoid extremely low levels of oxygen and/or high levels of carbon dioxide. Various predictive models of respiration rate of minimally processed food have been developed to optimize the packaging characteristics and/or the optimal gas composition. The idea behind designing packaging is to select materials, headspace conditions and package dimensions to provide adequate protection to the packaged food. The packaging design today involves strategies and solutions of products. Planning in a strategic way means to understand in which way innovation can be part of company target.

In: Food Science Research Summaries. Volume 3
Editor: Lucille Monaco Cacioppo
ISBN: 978-1-63117-864-1

Chapter 30

CHARACTERIZATION OF PLA, PCL AND SODIUM CASEINATE ACTIVE BIO-FILMS FOR FOOD PACKAGING APPLICATIONS

***M. Ramos*[1], *M. P. Arrieta*[1,2], *A. Beltran*[1] *and M. C. Garrigós*[1]**
[1]Analytical Chemistry, Nutrition & Food Sciences Department, University of Alicante, Alicante, Spain
[2]Mechanical and Materials Engineering Department, Polytechnic University of Valencia, Alcoy, Alicante, Spain

RESEARCH SUMMARY

Biodegradable polymers, such as poly(lactic acid) (PLA), polycaprolactones (PCL) and edible films obtained from caseinates exhibit an excellent equilibrium of properties and they have been recently proposed as potential substitutes of some conventional synthetic polymers. Antioxidant and/or antimicrobial active packaging is one of the most updated tendencies in research for food industry to extend shelf-life of packaged products during storage; where active agents are usually obtained from natural compounds such as vegetables and essential oils. In this sense, carvacrol and thymol (present in oregano and thyme essential oils) and hydroxytyrosol (3,4-dihydroxyphenylethanol, phenolic compound present in olive fruits) are well known for their antimicrobial and antioxidant activities, respectively. In this chapter, different active bio-films were studied: PLA with thymol, PCL with hydroxytyrosol, and sodium caseinate plasticized with glycerol and carvacrol and hydroxytyrosol as active agents. In all cases, 10 wt % of active agents were used. Two organo modified montmorillonites were also included for biopolyesters at 2.5 wt % (DELLITE 43B for PLA and CLOISITE 30B for PCL) in order to improve mechanical and barrier properties. A complete morphological, thermal, mechanical and functional characterization of all formulations was carried out. SEM and TEM micrographs showed homogeneous surface films in all cases. Thermal stability by TGA; oxygen transmission rate (OTR), color parameters and tensile tests were also evaluated. X-ray analysis was carried out in order to evaluate the incorporation of the nanoclays, showing a good interaction with both polymeric matrices. The obtained results showed a great potential of the obtained formulations for their incorporation in active food packaging.

In: Food Science Research Summaries. Volume 3
Editor: Lucille Monaco Cacioppo

ISBN: 978-1-63117-864-1

Chapter 31

METAL-BASED MICRO AND NANO-COMPOSITES AS ANTIMICROBIALS IN FOOD PACKAGING

***Avelina Fernandez*[*,1], *Elsa Lloret*[2], *Amparo Llorens*[1] and *Pierre Picouet*[2]**

[1]Instituto de Agroquímica y Tecnología de Alimentos, CSIC, Paterna, Valencia, Spain

[2]Departament de Tecnologia dels Aliments, Institut de Recerca i Tecnologia Agroalimentàries (IRTA), Monells, Girona, Spain

RESEARCH SUMMARY

Advanced organic/inorganic hybrid materials with tailored properties open interesting possibilities in food packaging and surface engineering. The resulting interactions of polymer matrices and inorganic nanomaterials, such as ceramic particles or metals, modify the mechanical, thermal and barrier properties of the polymers and in some particular cases provide antimicrobial activity. In this chapter, our results concerning the incorporation of metals in biodegradable polymers, exchanged in zeolites or as nanoparticles, and their effectiveness as antimicrobial agents are summarized. Nanoparticle migration and antimicrobial activity need to be critically analyzed in order to validate those innovative materials in future food packaging applications.

[*] Tel.: +34 963900022; Fax: +34 963636301; Email address: avelina.fernandez@iata.csic.es.

In: Food Science Research Summaries. Volume 3
Editor: Lucille Monaco Cacioppo
ISBN: 978-1-63117-864-1

Chapter 32

ALTERNATIVE TECHNIQUE FOR THE QUANTITATIVE DETERMINATION OF ANTIMICROBIAL ACTIVITY OF ANTIMICROBIAL PACKAGING FILM CONTAINING LIPOPHILIC MATERIALS

Panuwat Suppakul
Kasetsart University, Thailand

RESEARCH SUMMARY

Nowadays, many researchers of antimicrobial polymeric packaging films containing lipophilic materials face with a failure of antimicrobial effectiveness in films tested using a conventional assay e.g., agar disc diffusion. This assay, however, may lead to a misunderstanding of the properties of lipophilic antimicrobial packaging films. An alternative techniques termed, a vapour diffusion assay, is more reliable, repeatable, quantitative in determining the antimicrobial effectiveness of lipophilic antimicrobial packaging films. Hence, lipophilic antimicrobial packaging films are benefited by this alternative novel test method by overcoming the measurement shortcomings of the past. This chapter outlines these problem areas, methodologies and present data using a new test technique.

In: Food Science Research Summaries. Volume 3
Editor: Lucille Monaco Cacioppo
ISBN: 978-1-63117-864-1

Chapter 33

ANTIMICROBIAL PACKAGING: LAB DEVELOPMENTS AND POSSIBLE COMMERCIALIZATION

Kay Cooksey[1], Angela Richard[1] and Sunil Mangalassary[2]

[1]Department of Food Nutrition and Packaging Science,
Clemson University, SC, US
Department of Kinesiology and Nutritional Sciences,
California State University, Los Angeles, CA, US

RESEARCH SUMMARY

Methods to improve the safety of food and prevent foodborne outbreaks are continually being developed. Antimicrobial packaging is one of the methods being developed which are not meant to be a replacement for good manufacturing and sanitation practices but rather an additional hurdle against microbial growth. While antimicrobial additives may be directly added to foods, there is a growing trend to reduce direct additives in foods by incorporating them into packaging. Many of the additives added to food packaging will still need to meet 21 Code of Federal Regulations (CFR) for food additives and/or food contact notification (Ettinger, 2002). The main work has focused on selection of natural additives to be incorporated into a packaging system, screening for effective levels, incorporation into materials and methods to control release.

Nisin, a polypeptide, is a bacteriocin produced by *Lactococcus lactis* subsp. *lactis* (Jin et al. 2009). Since its discovery in the early 1900s, nisin has proven to be an effective inhibitor of Gram-positive bacteria (Bower, et al. 1995) and is now approved for use in 57 countries around the world and has been affirmed as generally recognized as safe (GRAS) in the United States (Jin, et al. 2009). Nisin is considered nontoxic and nonantigenic to humans because it is inactivated by proteolytic enzymes in the digestive tract, and assumed to be safe for use as a food preservative (Bower, et al. 1995). The lantibiotic nisin is used in a variety of commercially produced foods, including dairy products, meats, fish, and eggs because of its non-toxic qualities, heat stability, and inert behavior which does not create any off-flavors.

(Le Blay, et al. 2007; Schillinger et al., 1996). While nisin is primarily active against Gram positive bacteria including *Listeria* sp., *Staphylococcus*, *Clostridium*, and *Bacillus*, they are ineffective when used alone against Gram negative bacteria because of its inability to penetrate the outer membrane present in the cell wall of Gram negative organisms (Helander and Mattila –Sandholm, 2000). But studies have found that when used in combination with metal chelating agents like EDTA, nisin was able to exert antibacterial action against Gram negative organisms (Boziaris and Adams, 1999, Cutter and Siragusa, 1995). Many research studies have investigated nisin in combination with other antimicrobials, however, this review will only look at the efficacy of nisin against Gram-positive bacteria alone so that the methods can be compared.

What proves interesting is how the results of these methods can be different depending on the assay method that the research uses. The literature provides examples of the diversity in the varying assays including methods such as, the spot on lawn assay, agar diffusion assay and Kirby-Bauer method. There are also examples of the many methods to incorporate antimicrobials like nisin, into food packaging materials, including coating, the use of heat press, the use of adsorbent materials, and the commercial favored extrusion. There are also methods for testing the effectiveness in packaging material, including the film on lawn assay, which sometimes employs an agar overlay method.

In: Food Science Research Summaries. Volume 3
Editor: Lucille Monaco Cacioppo ISBN: 978-1-63117-864-1

Chapter 34

EVALUATION OF SAFETY OF RECYCLED PAPER AND BOARD FOOD CONTACT MATERIALS

Panagiotis G. Demertzis*
University of Ioannina, Department of Chemistry,
Laboratory of Food Chemistry, Ioannina, Greece

RESEARCH SUMMARY

Paperboard packages represent a large and constantly growing part of the food packaging industry, due to their advantages compared with other traditional packaging materials. It represents the largest section of the food and drink packaging market, being used in approximately the 50 % of all packaging. Moreover, half of the European paper and board production is directly or indirectly concerned by food-contact standards.

To protect the environment, new ways of recycling waste, including packaging materials (paper and plastics), are appearing on the market. The environmental pressure and the general tendency of recycling pose recycled plastics and paperboard in the situation of being used as food contact materials. Paper and board partly or fully produced from recycled fibers is already being used in contact with certain foodstuffs in many countries in Europe.

Recover paper and board may vary in origin and could include paper containing printing inks, adhesives, trace elements, waxes, fluorescent whitening agents and dyes, sizing agents, organochlorine substances, plasticizers, aromatic hydrocarbons, volatile organic compounds, curing and grease-proofing agents, amines, biocides and surfactants. Therefore, to ensure food safety, paper and board produced from recycled fibers can be used as food contact materials if it originates from specific quality of recovered paper and board which has been subjected to appropriate processing and cleaning, provided that the finished materials meet obligations laid out in EC Framework regulation 1935/2004. This regulation covers all materials and articles, which come in contact with food.

In the present chapter, a review on potential contaminants in recycled paper and board food contact materials is presented. A further objective was the investigation of the usability

* Tel: +30-26510-08340, Fax: +30-26510-08795, Email: pdemertz@uoi.gr.

of recycled paper and board for direct and/or indirect food contact applications by developing rapid analytical methods to test for various classes of organic pollutants that can be found as residues in recycled paper and boards and to test for migration levels of these compounds in foods and food stimulants. Obtained results were discussed in terms of the possibility from a safety point of view to use recycled fiber based materials for food contact applications.

In: Food Science Research Summaries. Volume 3
Editor: Lucille Monaco Cacioppo
ISBN: 978-1-63117-864-1

Chapter 35

FAKING AUTHENTICITY?

Maria Kniazeva*
Associate Professor of Marketing, School of Business Administration,
University of San Diego, San Diego, CA, US

RESEARCH SUMMARY

The research objective of this study is to analyze how marketers convey "faked" authenticity. Two questions inform this purpose: 1) what cues do marketers use in attempts to create an authentic image of the products through their packaging? and 2) are there any differences in the use of such cues by the literally authentic and literally inauthentic products?

This qualitative study centers on packaging narratives and utilizes food products claiming authentic content from various countries and sold in US stores. It is the author's proposition that personification of the product may enhance its perceived authenticity, and it is the "human touch" as opposed to the prevalent emphasis on an objective description of the product that augments the authenticity message.

* E-mail: kniazeva@sandiego.edu.

In: Food Science Research Summaries. Volume 3
Editor: Lucille Monaco Cacioppo
ISBN: 978-1-63117-864-1

Chapter 36

THE MULTISENSORY PACKAGING OF BEVERAGES

***Charles Spence*[*,1] *and Betina Piqueras-Fiszman*[1,2]**
[1]Crossmodal Research Laboratory,
Department of Experimental Psychology, University of Oxford,
South Parks Road, Oxford, England
[2]Department of Engineering Projects, Universitat Politècnica de València,
Valencia, Spain

RESEARCH SUMMARY

The last few years have seen the rapid development of a range of novel packaging technologies that are currently enabling many companies to engage in radical innovation in terms of the multisensory packaging of their food and drink products. Consequently, more and more companies are now starting to introduce new coatings and packaging formats to their product lines. But just what multisensory attributes should one's packaging have? In this review, we highlight some of the most important recent developments in multisensory packaging design and innovation, focusing specifically on the beverage sector. The empirical evidence demonstrating the importance of the sight, sound, feel, smell, and, in some cases, even the taste/flavour of a product's packaging is critically reviewed. The latest research demonstrates how, at its very best, multisensory beverage packaging can significantly enhance a consumer's multisensory product experience (we call this '*functional*' packaging). We also demonstrate how the latest insights from the fields of experimental psychology and cognitive neuroscience are now starting to provide insights concerning the design of novel multisensory packaging; Packaging that is starting to stimulate the consumer's senses more effectively and make the products (not to mention the experience of consuming them) more memorable and enjoyable.

* Correspondence to: Prof. Charles Spence, Crossmodal Research Laboratory, Department of Experimental Psychology, University of Oxford, South Parks Road, Oxford, OX1 3UD, U.K. TEL: +44-1865-271364; FAX: +44-1865-310447; E-mail: charles.spence@psy.ox.ac.uk.

In: Food Science Research Summaries. Volume 3
Editor: Lucille Monaco Cacioppo
ISBN: 978-1-63117-864-1

Chapter 37

Water Soluble Components in Green Teas and Their Effects on Human Health

Wenbiao Wu [*] ***and Mingliang Huang***
College of Food Science, Southwest University, Beibei, Chongqing, China

Research Summary

Green teas are widely consumed in the world. Its aqueous extract obtained by soaking in hot water is employed as a daily drink. In addition to water, its aqueous soluble components are taken into the body. So, these components mainly including polyphenols, xanthine alkaloids (caffeine, theobromine, theophyline, etc.), free amino acids (theanine, protein amino acids, etc.), saponins, minerals, etc., have been extensively discussed. The main context has put emphasis on the effects of methods of growing fresh leaf and producing teas on the concentration of these chemicals in teas, effects of soaking conditions on their solubility, their impacts on human health including beneficial effects, dietary safety, side effects, etc.

[*] To whom all correspondence should be addressed. E-mail address: wbwu62@yahoo.com.

In: Food Science Research Summaries. Volume 3
Editor: Lucille Monaco Cacioppo

ISBN: 978-1-63117-864-1

Chapter 38

GREEN TEA: VARIETIES, PRODUCTION AND HEALTH BENEFITS

***Nesreen Ali*[1], *Eman Roshdy*[1], *Mohamed Sabry*[2] and *Ayman Al-Hendy*[3,*]**

[1]Public Health Department, Faculty of Medicine, Sohag University, Egypt
[2]OB/GYN Department, Faculty of Medicine, Sohag University, Egypt
[3]Center for Women Health Research, Department of Obstetrics and Gynecology, Meharry Medical College, George Hubbard Hospital, Nashville, TN, US

RESEARCH SUMMARY

Tea is one of the most popular beverages consumed worldwide. Based on the manufacturing process, green, black, and oolong tea are the 3 major commercial types of tea. Green tea, without any fermentation, is processed to prevent the oxidation of green leaf polyphenols, although most polyphenols are oxidized in black tea or oolong tea during fermentation production. This fermentation converts catechin to theaflavins and thearubigins, consequently decreasing the catechin content. The polyphenols present in green tea are flavonols, commonly known as catechins, which contain 5 major subtypes: catechin, epicatechin, epicatechin gallate, epigallocatechin, and epigallocatechin gallate (EGCG). These natural compounds show diverse chemical and biologic activities, and are nontoxic under daily dose. In recent years, evidences from epidemiologic and animal studies have shown chemopreventive and anticancer potential of dietary polyphenols. Several studies have suggested positive correlations between human consumption of green tea and a lower incidence of gastric, esophageal, ovarian, pancreatic colorectal cancers. EGCG, the major polyphenol in green tea, was found in animal studies to inhibit carcinogenesis effectively and broadly in various organs, such as the esophagus, stomach, and duodenum. Some clinical studies used oral EGCG for extended periods of time, in which EGCG was very well tolerated with virtually no reportable side effects. In cardiac patients, EGCG is reported to have

*To whom all correspondence should be addressed. E-mail address: ahendy@mmc.edu.

improved endothelial function and increased brachial artery flow-mediated dilation that paralleled the changes in plasma EGCG concentration. Numerous studies have described effects of EGCG when used in overweight and obese individuals for extended durations. Our group has recently demonstrated the ability of EGCG to control the proliferation and induce apoptosis in human leiomyoma cells, *in vitro* and in animal models. Our preclinical data in fibroid animal models demonstrates that EGCG added to drinking water was able to induce a highly significant shrinkage of fibroid lesions compared to untreated controls. Overall, emerging evidence suggest green tea and its extract as safe treatment for several diseases.

In: Food Science Research Summaries. Volume 3
Editor: Lucille Monaco Cacioppo
ISBN: 978-1-63117-864-1

Chapter 39

Green Tea and Pancreatic Cancer Chemoprevention

Quan V. Vuong[1,*], Christopher J. Scarlett[1,2] and Paul D. Roach[1]

[1]School of Environmental and Life Sciences, University of Newcastle, Ourimbah, NSW, Australia

[2]Cancer Research Program, Garvan Institute of Medical Research, Darlinghurst, NSW, Australia

Research Summary

Green tea refers to non-oxidised tea and it is an abundant source of catechins, which have been found to have strong antioxidant activities. Recent epidemiological, *in vitro* and *in vivo* laboratory studies have linked catechins with the chemoprevention of certain types of cancers. In addition, green tea is also a rich source of theanine, which has also been linked with cancer chemoprevention and has been found to reduce the adverse effects of cancer chemotherapy with agents such as doxorubicin.

Among the various types of cancers, pancreatic cancer has received considerable attention lately because it is the fourth leading cause of cancer death and it has a dismal 5-year survival rate of less than 5%. In addition, pancreatic cancer has been found to be largely resistant to conventional therapeutic strategies. Several attempts have been made to study the link between green tea polyphenols and pancreatic cancer.

Therefore, this chapter describes the properties, putative roles and plausible mechanisms of action for the main green tea components in relation to cancer chemoprevention and reviews the recent findings linking green tea and pancreatic cancer chemoprevention.

[*]To whom all correspondence should be addressed. E-mail address: van.vuong@uon.edu.au.

In: Food Science Research Summaries. Volume 3
Editor: Lucille Monaco Cacioppo

ISBN: 978-1-63117-864-1

Chapter 40

EFFECTS OF GREEN TEA ON COGNITIVE FUNCTION AND MENTAL STATUS: EMERGING EVIDENCE

Wenbin Liang[1] and Andy H. Lee[2,*]
[1]National Drug Research Institute, Curtin University, Perth, WA, Australia
[2]School of Public Health, Curtin University, Perth, WA, Australia

RESEARCH SUMMARY

Green tea is usually prepared by adding boiling water to dried tea leaves, which are made from the plant *Camellia Sinensis*. Worldwide, tea is the second most popular beverage after water. Similar to alcohol and coffee, drinking of green tea can produce pleasant feelings. In other words, consumption of green tea is partly due to its biological effects on cognitive function and emotions. To date, extensive epidemiological, clinical and experimental studies have shown that green tea drinking is beneficial to many aspects of physical health. There is also emerging evidence suggesting that key compounds of green tea may promote mental status and health of the central nervous system. The most promising candidates are L-theanine and green tea catechins. This commentary reviews recent findings from experimental and epidemiological studies on the neurological effects of green tea, and discusses possible mechanisms of action.

[*]To whom all correspondence should be addressed. Phone: +61 8 92664180; Fax: +61 8 92662958; Email: Andy.Lee@curtin.edu.au.

In: Food Science Research Summaries. Volume 3
Editor: Lucille Monaco Cacioppo ISBN: 978-1-63117-864-1

Chapter 41

EFFECTS OF ANTI-OXIDATIVE STRESS OF GREEN TEA CATECHINS ON CARDIOVASCULAR DISEASES

Jun-ichi Suzuki[1]*, Mitsuaki Isobe[2], Yasunobu Hirata[1] and Ryozo Nagai[3]

[1] Department of Advanced Clinical Science and Therapeutics, University of Tokyo
[2] Department of Cardiovascular Medicine, Tokyo Medicaland Dental University
[3] Department of Cardiovascular Medicine, University of Tokyo

RESEARCH SUMMARY

Green tea catechins have many biological functions, including anti-oxidative, anti-inflammatory and anti-carcinogenic effects. Anti-oxidative effects of green tea catechins are characterized by the ability to inhibit free radical generation and scavenge free radicals. They also influence activation of transcription factors such as nuclear factor-kappa B, a multipotential promoter of inducible nitric oxide synthase and adhesion molecules. However, anti-oxidative effects of catechins on cardiovascular diseases have not yet been well investigated. In this article, we reviewed recent clinical and experimental papers to reveal the anti-oxidative effects of green tea catechins on cardiovascular diseases. In our laboratory, we administered green tea catechins into animal models of myocardial ischemia, myocarditis and cardiac transplantation to reveal the effects on the oxidative stress-induced pathological remodeling. From our results and other investigations, green tea catechins are potent agents for the treatment and prevention of oxidative stress-induced cardiovascular diseases because they are critically involved in the suppression of the stress.

*Department of Advanced Clinical Science and Therapeutics, Graduate School of Medicine, University of Tokyo, 7-3-1 Hongo, Bunkyo, Tokyo 113-8655, Japan; phone 81-3-5800-9116, fax 81-3-5800-9182, e-mail junichisuzuki-circ@umin.ac.jp.

In: Food Science Research Summaries. Volume 3
Editor: Lucille Monaco Cacioppo
ISBN: 978-1-63117-864-1

Chapter 42

ANTIDIABETIC PROPERTIES OF GREEN TEA CATECHINS

Yannis V. Simos*[*,1], Evangelia Tselikou[1], Katerina Touloupi[2], Ioannis Verginadis[1], Ilias Karagounis[1], Anastasia Velalopoulou[1] and Angelos Evangelou[1]

[1]Laboratory of Physiology, Faculty of Medicine, University Campus of Ioannina, Ioannina, Greece
[2]Laboratory of Pharmacology, Faculty of Medicine, University Campus of Ioannina, Ioannina, Greece

RESEARCH SUMMARY

Tea is the most consumed drink in the world after water. The medicinal value of tea has been historically well known and since ancient times, green tea has been considered as a healthy beverage. In recent years, green tea's potential health benefits have extensively been investigated. In the early years, scientists focused their research on the main polyphenolic group present in green tea, catechins, and their anticancer effects. Nowadays the health benefits of green tea catechins (GTC) are increasingly being explored against *uprising* diseases such as diabetes, metabolic syndrome and cardiovascular diseases. Epidemiological studies have indicated metabolic health benefits from regular consumption of green tea. Furthermore, cell and animal studies have provided data on the role of GTC in the regulation of glucose uptake from cells, insulin sensitivity or even regulation of beta cells function. Therefore, GTC health benefits led to the conclusion that green tea can be considered as a beverage with functional properties. However, although all the evidence from research on green tea are very promising, future studies are necessary to fully understand its contribution to human health. The aim of this chapter is to summarize the latest research (cell, animal, human and epidemiological studies) on the effects of GTC (epigallocatechin-gallate,

[*]To whom all correspondence should be addressed.E-mail: simosyannis@gmail.com.

epigallocatechin, epicatechin-gallate and epicatechin) against conditions accompanying diabetes pathology such as reduced glucose uptake, insulin resistance and beta cells function.

In: Food Science Research Summaries. Volume 3
Editor: Lucille Monaco Cacioppo
ISBN: 978-1-63117-864-1

Chapter 43

POLYPHENOL TEA: PROTEIN PROTECTION, CANCER PREVENTION AND NANOTECHNOLOGY

Kouass Sahbani Saloua*[*,1,2], *Kouass Salah,*[3]
Aloui Amine*[1], *Abbas Chiraz*[1], *El May Alya*[1] *and Landoulsi Ahmed*[1]
[1]Unité de recherche de Biochimie des lipides et interactions avec les macromolécules. Laboratoire de Biochimie et de Biologie moléculaire, Faculté des sciences de Bizerte, Zarzouna, Tunisia
[2]Groupe en Sciences des Radiations, Département de Médecine Nucléaire et de Radiobiology, Faculté de Médecine, Université de Sherbrooke,Québec, Canada
[3]Laboratoire de Cristallochimie et Matériaux, Faculté des Sciences de Tunis, Tunisia

RESEARCH SUMMARY

Green tea is rich in polyphenolic compounds, with catechins as its major component. Studies have shown that catechins possess diverse pharmacological properties that include anti-oxidative, anti-inflammatory, anti-carcinogenic, anti-arteriosclerotic and anti-bacterial effects. They found microwave-assisted extraction (MAE) increases polyphenol tea extraction and gamma irradiation increases its biological activity. Also, they suggest that hot water extracts were more effective in antioxidant activity. However, cold water extracts were more effective in scavenging ability on DPPH and hydroxyl radicals, and chelating ability on ferrous ions. The pure catechins and phenolic acids found in tea are more powerful than the antioxidant vitamins C in an in vitro lipoprotein oxidation model. It was reported that phenolic compounds have an ability to interact with proteins, especially proline-rich proteins (such as caseins). Proline groups have a strong affinity for the hydroxyl (–OH) groups of catechins. Most of the literature suggests that hydrophobic and hydrogen bindings are responsible for polyphenol–protein interactions. Polyphenols may interact covalently or non-covalently with proteins. EGCG can bind to a protein cysteinyl thiol group through

[*]To whom correspondence should be addressed. Tel.: (001)819 346 1110 ext 12103 / (216) 23 347283; Fax: (001) 819-564-5442 /(216)-72-590566;E-mail: kouasssahbanisaloua@rocketmail.com.

autoxidation. The non-covalent interactions between phenolic compounds and proteins have been suggested to be created by hydrophobic interactions, which may subsequently be stabilized by hydrogen binding. It has been found that tea decreased the mammary gland tumor multiplicity and volume, and milk and tea had a greater protective action. Polyphenol tea has been shown to safely suppress proinflammatory pathways of chronic diseases including cancer; however, their low bioavailability in vivo limits their use in preventing and treating cancer. The utilization of encapsulated polyphenols, instead of free compounds, can effectively alleviate these deficiencies. Milk proteins have been used to encapsulate polyphenol.

In: Food Science Research Summaries. Volume 3
Editor: Lucille Monaco Cacioppo
ISBN: 978-1-63117-864-1

Chapter 44

GREEN TEA AND SKIN BENEFITS

***Neti Waranuch**[*,1] **and Wudtichai Wisuitiprot**[2]*
[1]Cosmetics and Natural Products Research Center, Faculty of Pharmaceutical Sciences, Naresuan University, Phitsanulok, Thailand
[2]Department of Thai Traditional Medicine, Sirindhorn College of Public Health, Phitsanulok, Thailand

RESEARCH SUMMARY

Most people accept that green tea is well-known as a medicinal plant. It is always suggested for daily health promotion. Moreover, its extracts have been widely used as topical applications for wound-healing, anti-aging, and disease treatments. Catechins are the prominent compounds in green tea extract and are promising ingredients in dermatological products. They are highly reactive with other compounds, such as reactive oxygen species and biologic macromolecules to scavenge free radicals or initiate biological effects. Although green tea presents an excellent result in *in vitro* studies with promising activity to benefit human health, the result of *in vivo* studies has not been fully satisfied. A few clinical studies reported that green tea extract significantly improved skin elasticity and increased the dermis's thickness; thus, providing skin wrinkle relief. The skin inflammation induced by UV radiation was decreased substantially when pre-treating the skin with green tea extract. Many researchers have attempted to explain the cause of skin improvement by several mechanisms. However, the complete picture has not been established. On the other hand, some studies revealed that topical formulation of green tea extract could not improve skin nourishing, and skin moisture tended to decrease when applying green tea products for a long time. The discrepancy of the *in vivo* study results is due to the limitation of catechins' properties. Instability, less skin permeation, and cutaneous metabolism play a crucial role in the effectiveness of green tea application. Although green tea's benefits for the skin seem unsatisfied in real life, a recent study revealed that the appropriate preparation of green tea formulation and delivery technology can diminish catechin limitations.

* To whom all correspondence should be addressed. E-mail:netiw@nu.ac.th.

In: Food Science Research Summaries. Volume 3
Editor: Lucille Monaco Cacioppo
ISBN: 978-1-63117-864-1

Chapter 45

NON-COVALENT INTERACTIONS OF GREEN TEA POLYPHENOL: SOURCE OF NOVEL COLLOIDAL STRUCTURES

Ashok R. Patel and Krassimir P. Velikov
Unilever R&D, Vlaardingen, The Netherlands

RESEARCH SUMMARY

Epigallocatechin gallate (EGCG) - the primary polyphenol found in green tea extract is one of the widely researched polyphenolic compounds in food industry due to its strong anti-oxidant properties. The propensity of EGCG to interact non-covalently with macromolecules like proteins and polymers has been well documented. Here we present the review of the formation of colloidal particles from non-covalent interactions of EGCG with cellulose derivative which could be further utilized to generate novel micro-structures for encapsulation applications.

In: Food Science Research Summaries. Volume 3
Editor: Lucille Monaco Cacioppo
ISBN: 978-1-63117-864-1

Chapter 46

CHINESE GREEN TEA: PRODUCTION, VARIETAL RESOURCES AND ADAPTABILITY

***La Li*[1] *and Chun Xiao*[1,2]**
[1,2]College of Food Science, Southwest University, Tian Sheng Qiao, Beibei, Chongqing, PRC

RESEARCH SUMMARY

An overview has been given, which associates with the development of the green tea processing as well as the cultivars resources and adaptability of the green tea in China. Chinese green tea cultivars are very ample and some researching progresses in their breeding and evolution have been made. These existing records of cultivars, strain, tea tree groups and big tea tree are still useful resources of developing or breeding superior varieties. Nevertheless, these areas still need to be further studied.

[1]To whom all correspondence should be addressed. E-mail: xndxxc@163.com.

In: Food Science Research Summaries. Volume 3
Editor: Lucille Monaco Cacioppo
ISBN: 978-1-63117-864-1

Chapter 47

PHYTOCHEMICALS AS POTENTIAL PHARMACOPHORES TO TARGET DIABETES

Mahomoodally Mohamad Fawzi*
Department of Health Sciences, Faculty of Science,
University of Mauritius, Reduit, Mauritius

RESEARCH SUMMARY

The sensible and liberal use of phytotherapy in the treatment and/or management of panoply of human diseases have always been supported by the ancient medicinal literature worldwide. Indeed, prior to the introduction of insulin therapy, the management of diabetes mellitus (DM) relied closely on dietary measures, which included the use of traditional plant therapies. The Papyrus Ebers of 1550 B.C. recommended a high-fiber diet of wheat grains for the management of diabetes. Currently more than 1200 species of organisms have been used ethnopharmacologically or clinically to treat/manage symptoms and complications associated to DM and panoply of reviews/reports of plants with known antidiabetic activity or traditional use as antidiabetic remedies have been published. In general, discussion pertaining to medicinal agents from plant tends to center around plant secondary metabolites or phytochemicals, i.e., non-ubiquitous constituents with no known essential role in the plant's metabolism. However, it has been postulated that bioactive plant secondary metabolites may play a pivotal role in chemical defense mechanisms and are likely molecules for the hypoglycemic agents in these plants. Recently, the growing interest of researchers in traditional medicine has led to discovery and establishment of biological activity of many potent phytochemicals possessing hypoglycemic activity (e.g., glycosides, lavonoides, terpenes, steroidal saponins, alkaloids, galactomannan gum, pepdidoglycans, polysaccharides, hydroglycans, guanidines, glycopeptides and flavonoids amongst others). A plethora of plant's derived bioactive compounds representing numerous phytochemicals has demonstrated activity consistent with their use in the traditional treatment/management of DM. Even the discovery of the widely used hypoglycemic drug, metformin has been

* E-mail address: f.mahomoodally@uom.ac.mu; mmfawzi@gmail.com.

developed from the traditional approach of using *Galega officinalis*, and which subsequently led to the synthesis of biguanides. Thus, phytochemical could be a potential source of anti-diabetic pharmacophores and template molecules geared towards alleviating chronic complications of diabetes. This present book chapter endeavors to present an overview on how phytochemicals, particularly from common traditionally used medicinal plants, may be helpful in glucose homeostasis thereby reducing the risk of DM and related complications.

In: Food Science Research Summaries. Volume 3
Editor: Lucille Monaco Cacioppo

ISBN: 978-1-63117-864-1

Chapter 48

PHYTOCHEMICALS AND THEIR POTENTIAL EFFECTS ON INFLAMMATORY BOWEL DISEASE

Susana Gorzalczany*
Pharmacology Department, Facultad de Farmacia y Bioquímica,
Universidad of Buenos Aires, Argentina

RESEARCH SUMMARY

The incidence of gastrointestinal diseases has become a public health concern with high social costs. Inflammatory bowel disease (IBD), which includes ulcerative colitis and Crohn's disease, is a chronic intestinal disorder. Their etiology is at present unknown and the therapy remains palliative rather than curative, exhibiting distressing side effects. There is a clear demand for new agents for the treatment of IBD. The interest of isolated compounds from natural sources as potential drugs for the treatment of IBD is growing in the last years. Phytochemicals include different kinds of chemical structures. Many molecules showed antioxidant activity, anti-inflammatory effect and immumodulatory properties and might be useful as adjunct therapy for IBD. The effect of phytochemicals in the complex cell-signaling pathway of gastrointestinal diseases will be discussed and their mechanism of action analyzed.

* E-mail address: sgorza@ffyb.uba.ar.

In: Food Science Research Summaries. Volume 3
Editor: Lucille Monaco Cacioppo

ISBN: 978-1-63117-864-1

Chapter 49

BIOACTIVE COMPOUNDS IN WHOLE GRAIN CEREALS

Maria Teresa Pedrosa Silva Clerici[1*]
and Elizabeth Harumi Nabeshima[2]
[1]College of Nutrition, Federal University of Alfenas – UNIFAL – MG, Brazil,
[2] Institute of Food Technology (ITAL), Brazil

RESEARCH SUMMARY

Cereals are edible portions of the grass family, *Gramineae*, whose grains and products are an important source of energy, carbohydrate, protein and fiber, containing a range of micronutrients such as vitamin E, some of the B vitamins, magnesium, and zinc. Currently, it has been recognized as an important source of phytochemical compounds. The final nutrient content of a cereal after milling will depend on the extent to which the outer bran and aleurone layers are removed, as this is where the fiber, vitamins, minerals and bioactive compounds tend to be concentrated. Epidemiological studies have shown that regular consumption of whole grain cereal products is associated with reduced development of chronic diseases, such as cardiovascular diseases, type 2 diabetes, and some kinds of cancers. Phenolic acids and flavonoids represent the most common form of phenolic compounds found in whole grains, and they are among the major and most complex groups of phytochemicals. The increased use of whole grain flour in breads, pastas and other products can, in addition to promoting nutritional benefit, promote a better exploitation of cereal and reduce the by-product formation. This chapter aims to show studies regarding the great potential of use of whole grain cereals as a source of bioactive compounds.

* E-mail address: mariateresapedrosa@gmail.com.

In: Food Science Research Summaries. Volume 3
Editor: Lucille Monaco Cacioppo

ISBN: 978-1-63117-864-1

Chapter 50

ANTIOXIDANT AND ANTIMICROBIAL ACTIVITIES OF BERRIES EXTRACTS OF FIVE *JUNIPERUS* SPECIES UNDER *JUNIPERUS* SECTION FROM TURKEY

Maria Fernanda Taviano**, Andreana Marino, Ada Trovato, Rita De Pasquale and Natalizia Miceli
Pharmaco-Biological Department, University of Messina, Messina, Italy

RESEARCH SUMMARY

In Turkish folk medicine numerous *Juniperus* species, particularly those under the *Juniperus* section, are frequently employed to treat several diseases.

This work was designed to evaluate the antioxidant and antimicrobial potential of methanol berries extracts of *Juniperus* species under *Juniperus* section from Turkey: *Juniperus communis* L. var. *communis*, *Juniperuscommunis* L. var. *saxatilis* Pall., *Juniperus drupacea* Labill., *Juniperus oxycedrus* L. subsp. *oxycedrus*, *Juniperus oxycedrus* L. subsp. *macrocarpa* (Sibth. & Sm.) Ball.

The total phenolic content of *Juniperus* spp. extracts, evaluated by Folin-Ciocalteau method, ranged from 5.14 ± 0.06 mg GAE/g extract (*Joo*) to 59.17 ± 1.65 mg GAE/g extract (*Jcc*), and it decreased in the following order: *Jcc*>*Jd*>*Jom* = *Jcs*>*Joo*.

The antioxidant properties of the extracts were examined in different *in vitro* systems. *Jd* showed the best activity in the DPPH (1,1-diphenyl-2-picrylhydrazyl) test (IC_{50}=0.38 ± 0.02 mg/mL), in the Fe^{2+} chelating assay (IC_{50}=2.26 ± 0.06 mg/mL), and in the TBA (ThiobarbituricAcid)test (IC_{50}=2.47 ± 1.13 µg/mL); *Joo* resulted in the most active extract (4.05 ± 0.09 ASE/mL) in the reducing power assay. In addition, the antioxidant activity of the extracts was correlated to their phenolic content. The obtained results indicated that *Juniperus* spp. extracts possess both primary and secondary antioxidant activity; moreover, phenolic compounds appear to be responsible for the primary antioxidant effects.

* E-mail address: mtaviano@unime.it.

Since *J. communis* berries are traditionally used for their antiseptic action, it seemed interesting to evaluate the antimicrobial potential of all *Juniperus* spp. included in *Juniperus* section. The antimicrobial activity of *Juniperus* spp. extracts against bacteria and yeasts was evaluated by standard methods. The efficacy was appreciable on gram-positive only. Among the extracts *Jd* showed the higher bacteriostatic activity (*MIC*: 78.12-312.50 μg/mL) than *Jcs* and *Jcc* (*MIC*: 156.25-1250.00 μg/mL) followed by *Joo* and *Jom*(*MIC*: 625.00-1250.00 μg/mL). The *Staphylococcus aureus* was the most susceptible strain tested (*MIC*: 78.12-625.00 μg/mL).

The results give a scientific basis to the traditional utilization of these *Juniperus* species, also demonstrating their potential as sources of natural antioxidant and antimicrobial compounds.

In: Food Science Research Summaries. Volume 3
Editor: Lucille Monaco Cacioppo
ISBN: 978-1-63117-864-1

Chapter 51

Phenolic Compounds, Health Effects and Antioxidant Properties of Berries

***Aleksandra Konić-Ristić*[1]*, Katarina Šavikin*[2,*]*, Gordana Zdunić*[2]*, Dejan Gođevac*[3]*, Teodora Janković*[2] *and Nebojša Menković*[2]**

[1]Institute for Medical Research, University of Belgrade, Belgrade, Serbia
[2]Institute for Medicinal Plant Research "Dr. Josif Pančić", Belgrade, Serbia
[3]Institute for Chemistry, Technology and Metallurgy, University of Belgrade, Belgrade, Serbia

Research Summary

The biological value of berries relies on their nutritive and non-nutritive constituents. They are good sources of vitamins (vitamin A, C, E, folic acid), pro-vitamins and related compounds (carotene and lutein), minerals (calcium and selenium) and phytosterols, but the proposed health benefit is usually connected with phenolics as the main biologically active compound of berries. Berry phenolics include anthocyanins, flavonols, flavanols, condensed tannins (proanthocyanidins), hidrolyzable tannins (ellagitannins and gallotannins) and phenolic acids. Berries are recognized as the best dietary sources of anthocyanins. Different berries are distinguished by the presence of different classes of phenolics or compounds within the same class, their content and chemical structure (substituents, glycosylation, conjugation), potentially influencing their physiological properties and health effects. Analytical assessment of the phytochemical content of berries, the effects of storage and processing and putative interaction with other dietary compounds is essential for evaluation of their health benefit. It is of great importance to emphasize the biodiversity in the context of different phytochemical content in wild versus cultivated species, different varieties of the same species, or influenced by different cultivating conditions. Beneficial effects of berries consumption in health promotion and prevention of chronic diseases, such as cancer or

* Email address : ksavikin@mocbilja.rs.

cardiovascular diseases are shown in numerous epidemiological studies. In *in vitro* and animals studies berry fruits, their extracts and isolated compounds showed antioxidant, anticancer, antidiabetic, hypoholesterolemic, anti-inflammatory, vasoprotective, cardioprotecive, antiplatelet, antitrombotic or antimicrobial effects. Some of these effects are also shown in clinical trials in humans. Phenolic compounds of berries and their extracts have strong antioxidant potential that is often used to explain their health-promoting properties. Recent studies postulated that their antioxidant potential, as a basic mechanism of biological activity, is compromised by their low bioavailability. Consequently further research in this area is targeted towards other mechanisms of action, the evaluation of main sites of action and further elucidation of potential of berry phenolic compounds as antioxidants, based on their structure or metabolism, both *in vitro* and within the human body.

In: Food Science Research Summaries. Volume 3
Editor: Lucille Monaco Cacioppo
ISBN: 978-1-63117-864-1

Chapter 52

OVERVIEW OF THE CHEMICAL, NATURAL OCCURRENCE AND ANTIOXIDANT ACTIVITY OF TERPENES

E. González-Burgos and M. P. Gómez-Serranillos*
Department of Pharmacology, Faculty of Pharmacy.
Universidad Complutense de Madrid (UCM), Madrid, Spain

RESEARCH SUMMARY

Terpenes are one of the most abundant groups of natural products that occur mostly in plants, being identified more than 30,000 different compounds. Chemically, terpene group is one of the most diverse in structure, sharing isoprene units as a common structural motif. The number of isoprene units that defined each terpene compound classifies them into several groups including hemiterpenes, monoterpenes, sesquiterpenes, diterpenes, sesterpenes, triterpenes, tetraterpenes and politerpenes.

Terpenes, as a natural product group, have sparked a great interest in research because of their relevant and broad spectrum of health-promoting effects such as anti-inflammatory, antimicrobial, antitumor and antioxidant, among others. Nowadays, new findings in pharmacological area are in constant growing.

Since oxidative stress has been involved in the pathophysiological development of approximately one hundred different diseases including cancer, autoimmune diseases, heart attack and neurodegenerative diseases, there have been numerous attempts to cope with this redox imbalance. Exogenous compounds with antioxidant capacity provide a protective effect against free radical overproduction and its consequence cellular damage, being this one the most promising therapeutic strategy. Over the last recent years, there is an increasing interest in the search of antioxidant compounds among terpenes, existing emerging number of reports focused on this issue.

---

* E-mail address: pserra@farm.ucm.es Dra. M. Pilar Gómez-Serranillos. Full Professor, Department of Pharmacology, Faculty of Pharmacy. Universidad Complutense de Madrid (UCM), Madrid Spain.

This chapter will cover the chemistry, the natural occurrence and the most recent findings in antioxidant activity of terpenes.

In: Food Science Research Summaries. Volume 3
Editor: Lucille Monaco Cacioppo
ISBN: 978-1-63117-864-1

Chapter 53

SUPERCRITICAL EXTRACTION OF PHLOROGLUCINOLS FROM ST. JOHN'S WORT (*HYPERICUM PERFORATUM* L.) – PROCESS OPTIMIZATION, MATHEMATICAL MODELING AND ANTIMICROBIAL ACTIVITY

Sandra B. Glisic*[*] *and Aleksandar M. Orlovic
Faculty of Technology and Metallurgy,
University of Belgrade, Belgrade, Republic Serbia

RESEARCH SUMMARY

St. John's Wort (*Hypericum perforatum* L.) has been widely used as an anti-inflammatory and healing agent in traditional medicine. Some para-pharmaceutical products, prepared from the extract of St. John's Wort, are frequently used because of their anti-depressive and antiviral properties. Moreover, the extract prepared from St. John's Wort (*H. perforatum* L.) has been also reported to have an antimicrobial, anti-inflammatory, and healing effect.

Thus, extraction of a specific component from St. John's Wort which gives a high yield of total extract and high content of desired pharmacological active compounds is obviously in focus of many research activities today. Despite the evident clinical effects, there is still a controversy regarding the active principle of the extracts. Some authors attribute the antidepressant activity to hypericin and its derivatives, whereas, other studies are in favor of hyperforin.

Hyperforin, as a phloroglucinol derivative is a relatively lipophilic compound which is susceptible to rapid thermal, light-induced and oxidative decomposition. Adhyperforin presents its methyl derivative.

The supercritical fluid extraction (SFE) is an outstanding method for the extraction of natural agents and valuable components from plant materials, because heat-sensitive

[*] Phone: +381 11 3303707; Fax: +381 11 3370387; Email: sglisic@tmf.bg.ac.rs.

compounds can be extracted without any degradation, and in addition, it is an environmentally acceptable technology. Supercritical carbon dioxide ($SCCO_2$) is the most useful supercritical fluid in the food industry, because it is non-toxic, non-explosive, and available in high (food grade) purity and can be removed from the extracted products without leaving any residue.

The main goal of this study was oriented to the investigation of process optimization of the best extraction method applicable to intergradient in food and pharmaceutical industry – supercritical fluid extraction as well as the antimicrobial activity of obtained extracts. The first part of study will represent the literature review of St. John's Wort occurrence in nature, their health effects and medical and pharmaceutical use.

In: Food Science Research Summaries. Volume 3
Editor: Lucille Monaco Cacioppo
ISBN: 978-1-63117-864-1

Chapter 54

NATURAL ANTIOXIDANTS IN FOOD

Gui-Fang Deng[1], Xiang-Rong Xu[2], Sha Li[1], Fang Li[1], Shan Wu[1] and Hua-Bin Li[1,*]
[1]Guangdong Provincial Key Laboratory of Food, Nutrition and Health, School of Public Health, Sun Yat-Sen University, Guangzhou, China
[2]Key Laboratory of Marine Bio-resources Sustainable Utilization, South China Sea. Institute of Oceanology, Chinese Academy of Sciences, Guangzhou, China

RESEARCH SUMMARY

The oxidative stress imposed by reactive oxygen species plays an important role in many chronic and degenerative diseases, such as cardiovascular diseases, cancer, diabetes mellitus, ageing and neurodegenerative diseases. The scavenging of reactive oxygen species is thought to be an effective measure to depress the level of oxidative stress of organism for prevention and treatment of some chronic and degenerative diseases. It has been reported that intake of vegetables and fruits was inversely associated with the risk of many chronic diseases, such as cardiovascular diseases and cancer.

Natural antioxidants in vegetables and fruits are considered to be responsible for these health benefits. In recent years, there is considerable interest in studying antioxidant activities of food. In this chapter, natural antioxidants in food are reviewed, and special attention is paid to antioxidant activities of vegetable, fruit and spice, including extraction methods of antioxidants, evaluation methods of antioxidant activity, antioxidant capacities and total phenolic contents, separation and identification of antioxidants as well as structure-activity relationship of antioxidant. The review provides an account on our research work combined with a reference of the literature published in the past ten years.

* Corresponding author. Tel.: +86-20-87332391; fax: +86-20-87330446. Email address: lihuabin@mail.sysu.edu.cn (H.-B. Li).

In: Food Science Research Summaries. Volume 3
Editor: Lucille Monaco Cacioppo
ISBN: 978-1-63117-864-1

Chapter 55

REGULATION OF INFLAMMATION BY DIETARY PHYTOCHEMICALS: CURRENT PERSPECTIVES

Marta González Castejón and Arantxa Rodriguez Casado*
IMDEA Food Institute, Madrid, Spain

RESEARCH SUMMARY

Acute respiratory distress syndrome, sarcoidosis, psoriasis, inflammatory bowel diseases, rheumatoid arthritis, osteoarthritis, and atherosclerosis are examples of inflammatory diseases. Also, chronic inflammation is associated to other serious chronic illnesses - including heart disease, obesity, diabetes, Alzheimer's disease, allergy, asthma, and many forms of cancer. Inflammation is a key feature of autoimmune diseases and other conditions that affect the immune system. Whether inflammation is the root cause of chronic illness, merely accompanies them, or occurs as a result of the disease, different dietary choices play significant roles in inflammation development. Intake of specific foods that influence the inflammatory process is an easy daily way to prevent and reduce long-term risks. In recent years, increasing attention has been paid to medicinal plants, due to many epidemiological and clinical studies that have demonstrated how a proper diet rich in micronutrients reduces chronic inflammation and, simultaneously, helps controlling and managing chronic health conditions. For example, high levels of cytokines cause many of the detrimental consequences of autoimmune disease. Many plants have effects on cytokines, e.g., interleukins, interferons, and chemokines, naturally-produced proteins that modulate the immune response and trigger inflammatory responses to toxins, injury, viruses, and bacteria. Some medicinal plants known to reduce inflammation by modulation of cytokines secretion include *Zingiber officinale* (ginger), *Allium sativum* (garlic), *Taraxacum officinale* (dandelion), *Echinacea purpurea* (purple coneflower), and *Silybum marianum* (milk thistle).

* E-mail address: arantxa.rodriguez@imdea.org.

Plants exert their beneficial effects through the additive or synergistic action of an immense variety of biologically-active, non-nutritive compounds known as phytochemicals acting at single or multiple target sites associated with physiological processes. Moreover, dietary spice phytochemicals, such as curcurmin from turmeric (*Curcuma longa*) and capsaicin from red pepper (*Capsicum baccatum*) are now emerging as agents that can not only prevent but may even treat inflammation.

Many edible plants with anti-inflammatory properties can be used as the sole therapy in inflammatory disease or as complementary of conventional treatments including corticosteroids, allowing patients to take smaller doses or shorter courses of drugs. In addition to influencing inflammation, phytochemicals might provide steady protective effects against different diseases accompanying the inflammatory response. This chapter assesses the anti-inflammatory effects of certain plants, as well as the functional benefits of notable groups of phytochemicals. In it we discuss how spice-derived phytochemicals suppress inflammatory pathways and what their molecular targets are.

In: Food Science Research Summaries. Volume 3
Editor: Lucille Monaco Cacioppo

ISBN: 978-1-63117-864-1

Chapter 56

ANTIOXIDATIVE PHYTOCHEMICALS AND SKIN CARE

Frédéric Bonté ***and Gallic Beauchef***
LVMH Recherche, Saint Jean de Braye, France

RESEARCH SUMMARY

Skin is one of the most important parts of the body and represents approximately two square meters in the adult. Day by day, skin undergoes internal and external aggressions. Daily emotional stress, ultraviolet radiations and abrupt temperature variations make it a target of choice for the oxidative stress. The biological targets are numerous: DNA, proteins, lipids mitochondria... During ageing or in case of strong aggressions, the cutaneous defense systems are overflowed, and the skin will develop wrinkles, spots or pathologies. How natural substances can increase the systems of natural detoxification and defence is one of the current scientific questions. This article proposes to focus on these targets and phytochemicals, especially the main antioxidant natural substances used in the dermatological care of the skin. The article also relates to the recent discoveries in this field. The main biological targets and the expected efficiency of antioxidants phytochemicals used in cosmetics are detailed.

* E-mail address: fredericbonte@research.lvmh-pc.com.

In: Food Science Research Summaries. Volume 3
Editor: Lucille Monaco Cacioppo

ISBN: 978-1-63117-864-1

Chapter 57

RESEARCH PROGRESS ON MEDICINAL PLANTS OF URTICA LINN. GENUS

Zhang Yanxiu, Wang Ying, Ju Zhihe and Guan Feng*

Key Laboratory of Chinese Materia Medica
(Heilongjiang University of Chinese Medicine), Ministry of Education, Harbin, China

RESEARCH SUMMARY

As types of traditional folk herbs and ethnic drugs, medicinal plants of the Urtica Linn genus, with a long history in China, have been widely regarded by people and research in the field of medicine were gradually carried on. Due to the unique curative effect, widespread distribution and abundant resources, broad development prospects are present with these medicinal plants. In the other countries, *U. dioica* L. subsp. dioica, *U. urens* L. and *U. cannabina* L., etc. can still be widely used for the treatment of prostate disease, psoriasis, allergic rhinitis, leukemia, diabetes, anemia and eczema in addition to rheumatism arthritis. In recent decades, a great number of studies on the phytology, chemical constituent, bioactivity and clinical applications on the Urtica Linn genus have been done worldwide. This review summarizes and analyzes the research results in these fields, mostly in China. It will provide the basics for further development and utilization of these important medicinal plants, not only for the pharmaceutical industry, but also for food, cosmetics and so on.

* E-mail address: traciiy-guan@hotmail.com.

In: Food Science Research Summaries. Volume 3
Editor: Lucille Monaco Cacioppo
ISBN: 978-1-63117-864-1

Chapter 58

RESEARCH AND DEVELOPMENT ON CHEMICAL CONSTITUENTS AND PHARMACOLOGY OF *ACANTHOPANAX* GENUS

Gaofeng Liu, Chao Zeng and Zhiping Lin
Department of Pharmacy, The Second Affiliated Hospital of Harbin Medical University, Harbin, Heilongjiang, China

RESEARCH SUMMARY

The *Acanthopanax* genus plants of *Araliaceae* include 37 species in the world. They are mainly distributed in China, Korea, Japan and the far-eastern region of Russia of northeastern Asia.

The root and stem bark of these plants, belonging to the liver meridian and kidney channel, are warm-natured, acrid flavor and bitter. They mainly contain triterpenoids, diterpenoids, lignans, flavonoids, polyacetylenes, phenylpropanoids and polysaccharides. *Acanthopanax* genus plants have been used for a long time as tonics and sedatives to nourish the liver and kidney, strengthen bones and muscle, stimulate circulation to end stasis, eliminate hydroncus, as well as having anti-tumor, anti-inflammatory, anti-fatigue, anti-oxidant, anti-diabetic, hepatoprotective effects, and immunomodulatory effects. They have shown effects on the brain, heart and nervous system, etc., and have been clinically used for the treatment of tumor, rheumatism, diabetes, hypertension, chronic bronchitis, ischemic heart disease, waist and knee aching and gastric ulcer. In recent decades, a great number of chemical, pharmacological and clinical studies on *Acanthopanax* genus have been carried out worldwide. The present review is an up-to-date and comprehensive analysis of the ecology, biology, chemistry, pharmacology and clinical applications of *Acanthopanax* genus. It will provide scientific references for further exploiting the bioactivity ingredients from the *Acanthopanax* genus, and provide a basis for further development and utilization of these plants in the food, cosmetic and pharmaceutical industry.

In: Food Science Research Summaries. Volume 3
Editor: Lucille Monaco Cacioppo

ISBN: 978-1-63117-864-1

Chapter 59

Simultaneous Determination of Five Triterpenoid Saponins in the Root of *Caulophyllum Robustum* Maxim by HPLC-ELSD

Guo-Yu Li[1], Yong-Gang Xia[2], Jun Liang[2], Yan-Hai Zhang[2] and Hai-Xue Kuang[2]*

[1]Pharmaceutical College, Harbin Medical University, Harbin, China
[2]Key Laboratory of Chinese Materia Medica (Heilongjiang University of Chinese Medicine), Ministry of Education, Harbin, China

Research Summary

A high-performance liquid chromatographic (HPLC) method coupled with evaporative light scattering detector was developed for the simultaneous quantification of five major bioactive constituents in the root of *Caulophyllum robustum* Maxim. The five constituents were separated on a Kromasil ODS column by gradient elution using. The validation of the method included recovery, linearity, accuracy and precision (intra- and inter-day variation). The range of recoveries of this method was 95.75–99.79%, with all the constituents showing good linearity ($r2 > 0.99$). The accuracy and precision were satisfactory, with the overall intra- and inter-day variation being less than 4%. The present method has been successfully applied for the determination of all five constituents in *C. robustum*. The study may be considered helpful to the quality control of *Caulophyllum robustum* Maxim.

In: Food Science Research Summaries. Volume 3
Editor: Lucille Monaco Cacioppo ISBN: 978-1-63117-864-1

Chapter 60

CHEMICAL STUDY AND MEDICAL APPLICATION OF SAPONINS FROM THE GENUS *ARALIA*

Hai-Xue Kuang**, *Yong-Gang Xia and Guang-Rui Mu
Key Laboratory of Chinese Materia Medica
Heilongjiang University of Chinese Medicine,
Ministry of Education, Harbin, China

RESEARCH SUMMARY

The genus of *Aralia* (*Araliaceae*), composing ca. 40 species, is mostly native to Asia, Oceania and North America. In these species, half are used mainly for medicinal purposes such as anti-tumor, adaptogen-like effect, effects on nervous system, effects on heart, anti-inflammatory activity, anti-diabetic activity and hepatoprotective effects. Saponins are a group of naturally occurring plant glycosides, characterized by their strong foam-forming properties in aqueous solution. Due to the great variability of their structures, saponins always display anti-tumorigenic effects through varieties of antitumor pathways. In addition, there are a large amount of saponins that still either remain to be trapped or be studied in detail by the medicinal chemists. In the past few years, a lot of triterpene saponins have been isolated from the leaves, bark, stem and tender shoot of Aralia genus. In recent years, with the growing concern on natural medicine, more and more attention was paid to Aralia species. In this article, we will summarize the progress on chemical structure of triterpene saponins and bioactivities of this genus from the literature until April, 2012.

* E-mail address: hxkuang@hotmail.com.

In: Food Science Research Summaries. Volume 3
Editor: Lucille Monaco Cacioppo ISBN: 978-1-63117-864-1

Chapter 61

CHEMICAL CONSTITUENTS AND BIOACTIVITIES OF PLANTS OF THE GENUS *HYDRANGEA*

Yang Chunjuan*[*] *and Xiao Yang
College of Pharmacy, Harbin Medical University, Harbin, China

RESEARCH SUMMARY

Hydrangea is one of the largest genera of Hydrangeaceae (Saxifragaceae), comprising about seventy-three species. They are distributed in eastern Asia, and eastern-north America, and extend southward to the tropic of both hemispheres. The plants of genus Hydrangea are always used as ornaments, flowers and a kind of natural medicine. They mainly contain isocoumarins, secoididoid glucosides, iridoid glucoside, sesquiterpene glucosides, triterpenes, flavonol glycosides, cyanogenic glycosides, alkaloids, coumarins, phenolic acids and their derivants. The isocoumarins are the principle constituents and their contents are higher than the other type constituents. The bioactivities of the crude extracts and the compounds of Hydrangea include antiallergic activity, immunomodulatory activity, inhibitory effect on histamine release, inhibitory effects on degranulation and release of cells, antimicrobial and antifungal activities, anti-diabetic activities, hepatoprotective and cholagoic activity, anti-malarial activities and antiulcer activities. The present review summarized the chemistry and pharmacological effects of the genus.

[*] E-mail address: chunjuanyang@126.com.

In: Food Science Research Summaries. Volume 3
Editor: Lucille Monaco Cacioppo

ISBN: 978-1-63117-864-1

Chapter 62

PHYTOCHEMICALS IN CANCER RESEARCH: A REVIEW OF THEIR BIOLOGICAL ACTIVITIES AND ANTICANCER POTENTIALS IN PRECLINICAL AND CLINICAL STUDIES

Daniel L. Pouliquen* and Marc Grégoire
Centre de Recherche en Cancérologie Nantes-Angers,
Institut de Recherche Thérapeutique de l'Université de Nantes
Inserm, Nantes, France
Université de Nantes, Nantes, France
CNRS, UM, Nantes, France

RESEARCH SUMMARY

Since the 1960s, the observation of geographical differences in the incidence of cancers has led epidemiologists to analyze their potential relationship with fruit and vegetable intakes. The capacity of a number of natural substances present in the diet to inhibit the neoplastic effects of chemical carcinogens has been demonstrated in laboratory animals. Many phytochemicals have been isolated from plants, their structures identified and their biological activities analyzed. In recent decades, some chemical classes have been the subject of extensive studies, *in vitro* and *in vivo*, and their mechanisms of action, now well documented, open up interesting prospects for cancer prevention and treatment in patients, either alone or as adjuncts to current cancer therapies. These compounds include isothiocyanates and indoles, organosulfur compounds and curcuminoids. Other potentially anticarcinogenic molecules naturally present in food may significantly reduce the incidence of specific cancers. These consist of secondary metabolites classified according to their chemical structure: alkaloids, acetogenins, stilbenoids, simple phenols, phenolic acids, polyphenols (flavonoids, lignans, proanthocyanidins, tannins, quinones), carotenoids, terpenoids, saponins, and others. For some of these, chemopreventive effects during the initiation, promotion and/or

* Email address: Daniel.Pouliquen@univ-nantes.fr.

progression steps of carcinogenesis have already been described. The diversity of traditional plant pharmacopeias throughout the world has also provided an invaluable source of ethnopharmacological information in the search for new molecules to be tested against experimental cancer models. Finally, as some of these phytochemicals are able to target multiple deregulated cellular signaling pathways, without toxic effects in normal cells, an interesting challenge for the future is to determine the optimal mode of combining their specific potentials.

In: Food Science Research Summaries. Volume 3
Editor: Lucille Monaco Cacioppo
ISBN: 978-1-63117-864-1

Chapter 63

MAMMARY GLANDS, AQUAPORINS AND MILK PRODUCTION

Ali Mobasheri*

Division of Veterinary Medicine, School of Veterinary Medicine and Science, Faculty of Medicine and Health Sciences, University of Nottingham, Sutton Bonington Campus, Leicestershire, UK

RESEARCH SUMMARY

The mammary gland is a specialized, enlarged sudoriferous or sweat (apocrine) gland that produces and secretes milk during lactation. Milk consists of simple sugars, lipids, proteins, vitamins and minerals dissolved in water, which accounts for up to 88% per unit volume of milk. The water content of milk will vary depending on the animal species under investigation and the physiological state of the lactating animal. Current knowledge suggests that water is secreted across the mammary epithelium in a transcellular manner, in response to an osmotic gradient produced largely by the lactose content of milk. Milk yield and quality are important criteria for the dairy industry. Despite the economic importance of milk yield little is known about the physiological mechanisms responsible for water transport in the bovine mammary gland. Recent studies suggest that several aquaporin proteins are present in the rodent, bovine and human mammary gland. Aquaporins play fundamental roles in water and small solute transport across epithelial and endothelial barriers. Immunohistochemical techniques have confirmed the presence of AQP1 and AQP3 water channels in rat, mouse, bovine and human mammary glands. Studies from our laboratory suggest that in addition to AQP1 and AQP3 the AQP4, AQP5 and AQP7 proteins are expressed in different epithelial and endothelial locations in the mammary gland. This chapter discusses the potential functional role of aquaporin water channels in the transport of water and small solutes across endothelial and epithelial barriers in the mammary gland and explores how these membrane proteins may be involved in milk production.

* Corresponding author: E-mail: ali.mobasheri@nottingham.ac.uk, Phone: +44-01159516449; Fax; +44-01159516440.

In: Food Science Research Summaries. Volume 3
Editor: Lucille Monaco Cacioppo

ISBN: 978-1-63117-864-1

Chapter 64

SCIENCE AND PSEUDO-SCIENCE OF MILK IMPLICATIONS FOR HUMAN HEALTH

*A. Nikkhah**
Department of Animal Sciences, Faculty of Agricultural Sciences,
University of Zanjan, Zanjan, Iran

RESEARCH SUMMARY

Milk production in ruminants is a bio-complex process that necessitates conversion and renovation of least available plant materials (e.g., cellulose) into most enriched available nutrients. This development involves pregastric rumen fermentation of plant cells by microbes and production of volatile fatty acids (VFA) amongst others as substrates for mammary milk lactose, fat and protein synthesis. As a result, milk contains numerous bioactive substances that function beyond solely their nutritive value. Essential amino acids, specialized protein such as casein, lactalbumins and globulins, peptides, nucleosides, nucleotides, unsaturated fatty acids such as conjugated linoleic acids, sphingomyelins, fat soluble vitamins and minerals of mainly calcium are principal examples. Milk sufficiency for neonatal brain, nervous and immune systems, and bone development and supportive tissue growth for even up to two years without a major need for alternative foods is an evolutionary proof for its irreplaceable role in human nutrition. However, the increasing concerns of cardiovascular disorders, hypertension, and related complexities in modern populations due to improper nutrition seem to have contributed to forming a fallacious public perception about milk in general and milk fat in particular as a possible risk factor. Despite several recent emphases on the functional nature of milk as a whole in improving health status in different populations, the confusion exists where education is suboptimal. Insightful education on the science of milk ought to accompany research evidence to enable the public to discern a pseudo-science that unconsciously disregards milk as a saturated fat animal food with serious risks to human health. It is reasoned that animal fat in the form of milk should not be considered biologically similar to animal fats supplied by non-milk sources. With limited

* E-mail: nikkhah@znu.ac.ir

saturated fat intake from non-milk sources, increased milk consumption could bear a multitude of positive impacts on health even with high fat content. Milk is a collection of bioactive substances with unique nutritional properties that synergistically optimize the health of mind and physics in different age groups.

In: Food Science Research Summaries. Volume 3
Editor: Lucille Monaco Cacioppo

ISBN: 978-1-63117-864-1

Chapter 65

RUMINANT MILK AND HUMAN WELLBEING: A MULTI-SPECIES REVIEW OF THE NATURAL RELATIONS

*A. Nikkhah**
Department of Animal Sciences, Faculty of Agricultural Sciences,
University of Zanjan, Zanjan, Iran

RESEARCH SUMMARY

Milk is unarguably the most nutritious bio-fluid in the nature. Ruminant milk provides an ideal food for humans of all ages. Thus, proper appreciation of milk nutritional value by the public requires thorough education of its biological descriptions in major ruminant species. The chapter assembles nutritional characteristics for humans of milk components and production by cattle, buffaloes, sheep, goats, and camels. Qualitative exclusivities and quantitative contributions of milk products from different ruminants to modern and post-modern human nutrition are discussed. Unique beneficial effects of different milk fats, proteins, peptides, minerals, vitamins, and other bioactive substances on optimal meeting of essential nutrients requirements and reducing the incidence of metabolic and cardiovascular complexities, cancers, traumas, and cognitive issues are also described. The advanced ongoing education and knowledge dissemination will be an obligation for unbiased milk natural health properties to be most deservingly realized by the new millennium's humans.

*E-mail: nikkhah@znu.ac.ir

In: Food Science Research Summaries. Volume 3
Editor: Lucille Monaco Cacioppo
ISBN: 978-1-63117-864-1

Chapter 66

COLOSTRAL AND LACTOGENIC MATERNAL IMMUNITY: HUMORAL AND CELLULAR FACTORS OF INDUCTION AND TRANSMISSION TO THE NEONATE

H. Salmon*

INRA, IASP, Lymphocyte et Immunité des Muqueuses, Nouzilly, France

RESEARCH SUMMARY

Colostrum and milk, secretions of mammary gland (MG) are the two components of the post-natal delivery of maternal immunity to the neonate. In monogastrics, sIgA is the predominant colostrum or milk immunoglobulins depending upon to the degree of prenatal Ig transfer, whereas in ruminant IgG1 predominate. In ungulate as swine and ruminant absence of prenatal Ig transfer is compensated for by IgG enriched colostrum. These immunoglobulins, typically absorbed within the first 36 hours of life, by entering in blood, provide the newborn with the antibodies that arose from antigenic stimulation of the mother's systemic immune system and sustain the systemic protection of the neonate against invasive pathogens. In contrast, the passive mucosal protection of neonate mammals is dependent on the continuous supply until weaning of maternally secretory dimeric IgA (Monogastric) and IgG1 (ruminants), the so-called lactogenic immunity.

The plasma cells (cIgA+ cells) of the mammary gland produces sIgA, which is, then secreted into the milk via the poly-Ig receptor of mammary epithelial cells. These antibodies are produced in response to intestinal and respiratory antigens, including pathogens and commensal organisms likely to be encountered by the neonate shortly after birth. Protection is also mediated by cellular immunity, which is transferred via maternal cells present in mammary secretions. The mechanisms underlying the various immunological links, that is cell (cIgA+ cells) trafficking from inductor sites (mucosal surfaces) to MG, involve hormonally regulated addressins and chemokines specific to these compartments. As the same pattern of adhesion molecules vascular adressin/ homing receptor- MadCAM-1/$\alpha 4\beta 7$ in one

* Email : Henri.Salmon@tours.inra.fr

hand, VCAM-1/α4β1 in another and CCL20 chemokine was observed in small gut and nasal mucosa indicate the existence of cellular link between upper respiratory tract and MG in addition to the cellular entero-mammary link. By comparison, absence of MadCAM-1 in ruminant MG, in agreement with the absence of cellular link with the intestinal immune system together with the low blood sIgA translocation to MG explain the low levels of IgA in bovine mammary secretions. The enhancement of colostrogenic immunity therefore depends on the stimulation of systemic immunity, whereas the enhancement of lactogenic immunity depends among other means of appropriate stimulation at induction sites, gut and upper respiratory tract. In addition, mammary secretions provide factors other than immunoglobulins that protect the neonate and regulate the development of mucosal immunity.

In: Food Science Research Summaries. Volume 3
Editor: Lucille Monaco Cacioppo
ISBN: 978-1-63117-864-1

Chapter 67

THE IMPORTANCE OF MILK FATTY ACIDS IN HUMAN NUTRITION

Rey Gutiérrez Tolentino[1*], Salvador Vega y León[1], Claudia Radilla Vázquez[2], María Radilla Vázquez[2], Samuel Coronel Nuñez[1]and Marta Coronado Herrera[1]

[1]Professor-Researcher of the Department of Agricultural and Animal Production. Universidad Autónoma Metropolitana- Xochimilco, Mexico City
[2]ConsultingMédica Sur Hospital, Puente de Piedra Toriello Guerra, Tlalpan, CP., Mexico City, Mexico

RESEARCH SUMMARY

Milk and dairy products are foods of great physiological functionality; however, this can be optimized by means of technological advances that make it possible to enhance their potentiality. Milk fat is perhaps the most complex of the edible fats, in which more than 400 different fatty acids (FA) have been detected, although 30 have been documented as majority, given that the rest are present in trace amounts, but which are nonetheless important from the functional viewpoint. Milk fat is constituted of approximately 25% monounsaturated fatty acids, 5 % polyunsaturated fatty acids and 70 % saturated fatty acids. Some of these FA have been shown to possess interesting properties for the health of individuals, especially butyric acid (C4:0), which intervenes in the elimination of cancer cells in the colon, and the vaccenic acids (trans-11. C18:1) and conjugated linoleic acid (CLA, C18:2) have anti-proliferative and anti-cancer actions, thus they have been added to foods as a preventive measure against cancer of the colon, rectum and prostate. There is sufficient scientific evidence that CLA participates in the inhibition of arteriosclerosis, increases bone mineralization and improves immunological functions. From the nutritional perspective, fatty acids naturally present in milk and dairy products are functional ingredients that play an important role in human health.

* E-mail: reygut@correo.xoc.uam.mx

In: Food Science Research Summaries. Volume 3
Editor: Lucille Monaco Cacioppo

ISBN: 978-1-63117-864-1

Chapter 68

INVESTIGATIONS ON THE GEOGRAPHICAL ORIGIN OF COW MILK AND COMPARISON OF THE LIPID COMPOSITION OF COW AND BUFFALO MILK BY MEANS OF TRADITIONAL AND INNOVATIVE PHYSICO-CHEMICAL ANALYSES

A. Sacco[*], *D. Sacco, G. Casiello, V. Mazzilli, A. Ventrella and F. Longobardi*
Dipartimento di Chimica, Università di Bari "A. Moro", Bari, Italy

RESEARCH SUMMARY

The quality of food products depends upon several factors such as pedoclimatic conditions, humidity level, and average temperature that characterize the geographic area where the foodstuff is obtained or produced. Since the geographic origin is widely considered as a valid criterion for quality assessment, the European Community set rules to confer the Protected Designation of Origin (PDO) label to peculiar local foodstuffs on the basis of their origin. In particular, the PDO label can be assigned only if the original raw materials are harvested, produced, and processed in the specific geographic region that gives the final food product its name. In the last decades, efforts have been directed towards exploring innovative and competitive methods to certificate the geographical origin, i.e., for the assessment of food quality and for the establishment of authenticity and traceability of food products. Several food products from the South Italian region of Apulia, are widely appreciated and considered as typical; among these, one important example is represented by the Apulian cow mozzarella, that is a fresh and stringy-textured cheese, whose authenticity could be proved by determining the geographical origin of milk.

In this chapter, traditional techniques, High Performance Ion Chromatography (HPIC), Inductively Coupled Plasma Atomic Emission Spectroscopy (ICP-AES), Nuclear Magnetic

[*] Email: antonio.sacco@chimica.uniba.it

Resonance (NMR) spectroscopy and Isotope Ratio Mass Spectrometry (IRMS) were used in combination with Multivariate Statistical Analysis (MSA) to determine a number of components and, hence, for discriminating the geographical origin of Apulian from foreign milk samples. Furthermore, the lipid fractions (FA) of raw Apulian cow and buffalo milk samples were analysed by using an innovative and powerful technique, i.e., ^{1}H NMR spectroscopy in combination with MSA, in order to search for a method to discriminate cow from buffalo milk, on the basis of the species. For the attribution of whole signals observed by ^{1}H NMR, 2D-NMR techniques, such as Correlated SpectroscopY (COSY), Heteronuclear Multiple Quantum Coherence (HMQC), and Heteronuclear Single Quantum Coerence (HSQC) were usefully employed; the quantitative determinations were achieved by integration of selected peaks.

In: Food Science Research Summaries. Volume 3
Editor: Lucille Monaco Cacioppo
ISBN: 978-1-63117-864-1

Chapter 69

ALTERNATIVE APPROACHES FOR THE PREVENTION OF BOVINE MASTITIS. PROBIOTICS, BIOACTIVE COMPOUNDS AND VACCINES

María Elena Fátima Nader-Macías[1*], Cristina Bogni[2], Fernando Juan Manuel Sesma[1], María Carolina Espeche[1], Matías Pellegrino[2], Lucila Saavedra[1] and Ignacio Frola[2]

[1] Cerela-Conicet, Chacabuco, Tucumán, Argentina
[2]Department of Microbiology and Immunology, Universidad Nacional de Rio Cuarto, Río Cuarto, Córdoba, Argentina

RESEARCH SUMMARY

Bovine mastitis (BM), both clinical and subclinical, is the most important infectious disease in the dairy farm, and basically affects the quality and the quantity of milk production. They are responsible for the major economic losses in the dairy farm due to treatment, veterinary service, animal replacement, discarded milk and mainly reduced milk production. The main causative microorganisms are *Staphylococcus aureus*, *Streptococcus agalactiae*, *Streptococcus dysgalactiae*, *Streptococcus uberis* and *Escherichia coli*. Even though, current practices as proper milking hygiene and reduced exposure to pathogens contributes to decrease the occurrence of the disease, the treatment for BM relies heavily on the use of antibiotics, both for prophylaxis and therapy. These treatments have proved to be frequently ineffective to control chronic mastitis, causing a persistent bacterial reservoir within a herd with recurrent infections and the culling of infected animal from the productive circuit. During the last decades, world tendency to limit the use of antibiotics in dairy cattle, has lead researchers toward the study of cows natural defence mechanisms in order to ensure their absence in dairy products with the aim of satisfy consumers demand for "organic products". At the same time, the increase of functional products or "health-preventing foods" consumption supports the need to increase some other aspects of the animal welfare and

* E-mail: fnader@cerela.org.ar

breeding. Alternatives approaches are being applied for the prevention of this disease and highly promoted, that includes the application of vaccines, the use of probiotics or beneficial microorganisms, their metabolic products as organic acids and hydrogen peroxide, antimicrobial peptides, and/or other more physiological treatments. This chapter reviews the last advances in this area, the rational to support each one of the preventive applications, together with the products available in the international market. Also, the clinical trials supporting the use of the different approaches are included, to encourage the advances and studies in this area. In this way, is possible to help to the control of mastitis, the improvement of the health status of the animals, and in this way in the milk production.

In: Food Science Research Summaries. Volume 3
Editor: Lucille Monaco Cacioppo
ISBN: 978-1-63117-864-1

Chapter 70

MILK PRODUCTION OF HOLSTEINS UNDER MEDITERRANEAN CONDITIONS: CASE OF THE TUNISIAN POPULATION

A. Ben Gara[1*], B. Jemmali[1], H. Hammami[2], H. Rouissi[1], M. Bouallegue[3] and B. Rekik[1]

[1]Ecole Supérieure d'Agriculture de Mateur, Mateur, Université Carthage, Tunisia
[2]Animal Science Unit, Gembloux Agricultural University, Gembloux, Belgium
[3]Faculté de Sciences de Tunis, Tunisia

RESEARCH SUMMARY

Continued selection for improved milk production has resulted in sizeable genetic gains in Holsteins within countries with leading dairy industries, e.g., North America and some European countries. Recognition of the Holstein as the top yielding breed led to the spread of its genes worldwide. The main breeding objective in Tunisia has long been milk yield and breeding strategies to improve milk production were based on semen and heifer imports. Actually, Holsteins account around 223000 cows of which only 10% are officially enrolled in the national milk recording system that dates back to the early 1970's. Studies using data recorded on the Tunisian Holstein population focused on milk production, reproductive performances, and genetic parameters. A cow spends roughly 3.22 years in the herd. Average national production level from all completed records was around 6220 kg of milk over a 305-d lactation period. Maximum yields were reached during third lactation where mean test- day yields were 20.7, 0.70, and 0.65 kg for milk, fat, and protein yields, respectively. Mean test-day somatic cell score (mean = 3.8 over the first three lactations) was relatively high and fertility measures were below satisfactory levels. In addition to annual and seasonal variations originating mainly from climatic changes, Holstein performances varied with herd ownership (state, cooperative or farmers' herds) and the region (from the North to the Centre of the country) that are determinant of the management level and production system (range of 305-d

* Email : rekik.boulbaba@iresa.agrinet.tn

mean milk yield by management level: 4,623 to 7,375 kg). Genetic parameter estimates mirrored management levels and were overall moderate in the range of those reported in the literature for populations in countries with emerging dairy industries. The lack of a stable genetic improvement policy coupled with a national routine genetic evaluation remains the main weakness limiting Holstein performances in the country.

In: Food Science Research Summaries. Volume 3
Editor: Lucille Monaco Cacioppo
ISBN: 978-1-63117-864-1

Chapter 71

FRESH EWE MILK PRODUCTION AND CEREAL NUTRITION: A PERIPARTAL INTERACTIVE MODEL OF GRAIN CHOICE AND LEVEL

A. Nikkhah* and M. Karam-Babaei
Department of Animal Sciences, Faculty of Agricultural Sciences,
University of Zanjan, Zanjan, Iran

RESEARCH SUMMARY

Periparturient metabolic imbalances continue to challenge profitable ruminant farming. Sheep as a docile, cooperative, economical and highly adaptable ruminant is increasingly regarded as an experimental model for development of strategies that can attenuate physiological shifts in nutrient metabolism. The objective was to determine effects of peripartal dietary cereal choice, level, and their interaction on sheep production and metabolism. Twenty Afshari×Merino ewes (80.3 ± 2.0 kg body weight) were used in a completely randomized design experiment from 24 days prepartum through 21 days postpartum. Ewes were kept indoor in individual boxes (1.5 × 2.5 m) and received once daily at 0900 h total mixed rations (TMR) containing either 1) higher or 2) lower levels of concentrates based on either 1) corn grain (CO) or 2) 50 to 50 combination of wheat + barley grains (WB). Dietary concentrate level and cereal choice were studied in a 2 × 2 factorial arrangement. Ewes were stepped into the postpartum diet after feeding two prepartum diets, one from 24 to 10 days followed by another fed until lambing. Dry matter based dietary forages included 3 parts of chopped alfalfa hay and 1 part of corn silage. Prepartal DMI numerically increased by feeding CO vs. WB. Feeding CO and not WB at higher vs. lower level improved postpartum DMI. Lambing DMI tended to increase with the higher vs. lower WB (1.59 vs. 1.37 kg/d). Feeding CO instead of WB and feeding both at lower vs. higher level increased fecal pH. Postpartal rumen pH decreased when the inclusion rate of WB (5.7 vs. 6.2, $P<0.05$) and not CO (6.3 vs. 6.3) was increased. Rumen propionate concentrations

* Email: nikkhah@znu.ac.ir

were decreased (20.4 vs. 18.9 mmol/L), and acetate levels (67 vs. 70 mmol/L) as well as acetate to propionate ration (3.3 vs. 3.7) were decreased by feeding higher amount of both grain choices. Colostrum yield and composition, peripartal urine pH, lamb weight at lambing and 21 days of age, and placental weight and expulsion time were unaffected. Milk yield (1.64 vs. 1.27 kg/d) and milk fat yield (99 vs. 81 g/d) were increased by increasing dietary levels of both grains. Plasma glucose increased by feeding higher vs. lower WB (57.6 vs. 52.2 mg/dL). Feeding CO instead of WB tended to reduce peripartal plasma NEFA concentration (0.25 vs. 0.28 mmol/L) and significantly increased insulin to NEFA ratios (2.47 vs. 1.77). Data for provide novel nutritional insights into independent and interactive effects of dietary cereal choice and level on periparturient sheep productive and metabolic responses. These serve as a peripartal model for high-producing ruminants (e.g., dairy cows) and non-ruminants.

In: Food Science Research Summaries. Volume 3
Editor: Lucille Monaco Cacioppo

ISBN: 978-1-63117-864-1

Chapter 72

FATTY ACID COMPOSITION OF MILK LIPIDS IN RESPONSE TO DIETARY FISH AND SAFFLOWER OILS

K. L. Jacobsen, S. Harris, E. J. DePeters *** and S. J. Taylor***
Department of Animal Science,
University of California, Davis, CA, US

RESEARCH SUMMARY

The effects of dietary fish and safflower oils were studied using eight lactating Holstein cows, 4 primiparous (P) and 4 multiparous (M), in a split-plot Latin square design. Cows were fed a basal diet composed of 41% sliced alfalfa hay and 58% concentrate ingredients to evaluate the effects of type and level of oil supplementation. Type of dietary oil was either 100% safflower oil (S) or 75% safflower oil + 25% fish oil (FS), fed at either 400 g/cow (L) or 800 g/cow daily (H). Yields of milk and milk components and dry matter intake were not affected by type or level of oil supplemented. Concentration of total *trans* fatty acids (FA) in milk fat triacylglycerol increased with the addition of fish oil and the higher level of oil. The concentration of C18:3 n3 decreased as the level of oil in the diet increased. Fish oil supplementation, particularly at 800g/day, increased concentrations of *cis*-9 *trans*-11 C18:2 and *trans*-11 C18:1. The high oil level increased concentrations of C20:5 and C22:5, but not C22:6, in milk fat. The phospholipid (PL) fraction was a rich source of polyunsaturated FA, and feeding diets rich in linoleic acid in conjunction with a source of marine oil increased the *cis*-9 *trans*-11 C18:2 (rumenic acid, RA) and *trans*-11 C18:1 (*trans* vaccenic acid, TVA) concentrations in PL. The FA compositions of plasma triglyceride (TG) and PL were not affected by parity, and few FA were altered by type or level of oil in TG. Concentration of C20:5 was not impacted by type or level of oil in the diet, and C22:6 was not measurable. Total TG in plasma was not affected by treatments. Plasma PL FA were affected by oil type; *trans*-11 C18:1, C20:5, and C22:6 increased, and C18:3 n3 decreased with addition of fish

* Corresponding author: ejdepeters@ucdavis.edu

oil, although oil level did not influence FA profile. There was no significant effect on plasma concentration of TG and non-esterified FA for either type or level of oil supplementation, possibly due to the high removal rate of TG by the mammary gland. Although a source of C20:5 and C22:6, dietary fish oil primarily increased the TVA and RA concentrations in milk fat, particularly in the PL fraction.

In: Food Science Research Summaries. Volume 3
Editor: Lucille Monaco Cacioppo
ISBN: 978-1-63117-864-1

Chapter 73

GENETIC FACTORS AND DAIRY-TECHNOLOGICAL VALORISATION OF MILK

***Andrea Summer*[1, 2*], *Massimo Malacarne*[1], *Paolo Formaggioni*[1], *Piero Franceschi*[1] *and Primo Mariani*[1, 2]**

[1]Department of Animal Production, Veterinary Biotechnologies, Food Quality and Safety, University of Parma, Parma, Italy
[2]MILC Centre, University of Parma, Parma, Italy

RESEARCH SUMMARY

Among the factors of variation, the genetic ones play a fundamental role in the determination of milk characteristics for cheesemaking. The genetic factors are more important the more the coagulation is of rennet type. In these conditions, the micelle structure of milk can totally express its intrinsic properties of genetic origin (e.g., polymorphism of caseins). The micelle maintains its structure almost unchanged in all the phases of the whole cheese making process, influencing markedly the rheological properties of the curd.

Role of the phenotypic combinations of caseins α_{s1}, β, k – In this part of the chapter the phenotypes (electrophoresis in starch-urea gel pH 8.6) of α_{s1}, β and k caseins of 222 individuals' milk samples from Italian Brown cows were determined. The aim was to estimate the effect of the composed phenotype on the physico-chemical and dairy-technological characteristics of milk. The statistical analysis put in evidence significant differences among the 5 considered phenotypes, regarding both the contents of some constituents and the rennet-coagulation characteristics

Comparison among genetic types – The aim of this part of the chapter was to study the physico-chemical, chemical and technological characteristics of milk of some genetic types present in the production area of the Parmigiano-Reggiano cheese, in comparison with those of the Italian Friesian cow milk. For this aim, the following four comparisons were carried out: Italian Brown *vs* Italian Friesian, Jersey *vs* Italian Friesian; Ayrshire *vs* Italian Friesian.

* Email: andrea.summer@unipr.it

The comparison between Italian Brown and Italian Friesian was made at the level of vat milk; while the other comparisons were carried out at the level of herd milk.

Dairy technological parameters of the Italian Brown cow milk in relation to types A and B of the β-lactoglobulin – The genetic type of the β-lactoglobulin in individual milk samples from Italian Brown cows was determined. The aim of this part of the chapter was to estimate the relationships between types of β-lactoglobulin and characteristics of milk. The β-lactoglobulin BB milk showed particularly interesting technological parameters, i.e., higher casein content and higher casein number.

In: Food Science Research Summaries. Volume 3
Editor: Lucille Monaco Cacioppo

ISBN: 978-1-63117-864-1

Chapter 74

ANIMAL WELFARE IN DAIRY OPERATIONS: A POSTMODERN VIEW

***Zenobia C. Y. Chan*[1] *and Wing-Fu Lai*[2]**
[1]School of Nursing, The Hong Kong Polytechnic University, Hong Kong Special Administrative Region, People's Republic of China
[2]Department of Chemistry, The University of Hong Kong, Hong Kong Special Administrative Region, People's Republic of China

RESEARCH SUMMARY

Consumption of milk products increased sharply in Asia, which is widely believed to potentially be a significant export market in the future. For this reason, Asia becomes a target of export expansion among countries having powerful dairy production (Dong, 2005). Under the prevailing moral discourse, animals are simply a cash cow and appear to be exploitable for human wants. Given this situation, so far extensive efforts have been devoted to optimize the dairy operations to increase the quality and quantity per yield, at the expense of animal welfares. By focusing on the context of milk production, this article aims to challenge, with a perspective of postmodernism, the existing monistic anthropocentric conception on the use of animals in agricultural practices. It is hoped that this article can raise the awareness of ethical animal use, and offer an insight into the development of a more moral environment in dairy production in which both animal welfare and industrial development could be secured.

[1] E-mail: hszchan@inet.polyu.edu.hk (Z. C. Y. Chan)
[2] E-mail: rori0610@graduate.hku.hk

In: Food Science Research Summaries. Volume 3
Editor: Lucille Monaco Cacioppo
ISBN: 978-1-63117-864-1

Chapter 75

SOY: NUTRITION, CONSUMPTION AND HEART HEALTH

Janice I. Harland[1,*], Elaine S. Krul[2], Ratna Mukherjea[2] and Kevin Maki[3],
[1]HarlandHall Associates, The Stables,
Poulton, Cirencester, England
[2]Solae LLC, St. Louis, MO, US
[3]Biofortis-Provident Clinical Research, Addison, IL, US

RESEARCH SUMMARY

Soyfoods have increased in popularity and number over the last two decades, partly as a result of the US Food and Drug Administration (FDA) permitting a health claim for soy protein that recognises its ability to lower blood cholesterol, first approved in 1999. A similar health claim was granted in the UK in 2002. Interestingly, the increase in soyfoods' popularity has been associated with an increase in controversy. Two main issues will be discussed here. First, the original health claim has been challenged based on the results of human clinical studies conducted after the date of approval. The rationale for the challenge was summed up in a report compiled by the American Heart Association. Their Soy Advisory publication concluded that in studies after 1999, soy protein only modestly lowered low density lipoprotein (LDL)-cholesterol (C) and did not exhibit a dose-response effect. Second, the increased consumption of soy has led to greater scrutiny of soy phytochemicals, and there has been some debate as to whether the cholesterol lowering activity of soy protein is attributable to the naturally occurring isoflavone content, or other mechanisms.

Numerous randomized clinical trials (RCTs) have been conducted with soy protein (with and without isoflavones) and soy isoflavones alone since 1999. Data from these studies have been pooled and a number of meta-analyses have been conducted since the original analysis published by Anderson in 1995. In the main these analyses have assessed the effects of soy

* Email: jan.harland@harlandhall.co.uk

protein on lipid levels, in particular LDL-C. These meta-analyses are reviewed in this chapter with the aim of evaluating the importance of factors such as isoflavone content, protein intake, baseline cholesterol concentration and various other factors on the lipid-altering effects of soyfoods consumption. Despite employing varying inclusion criteria, these analyses consistently report a reduction in LDL-C of ~0.2 mmol/L, equivalent to a reduction of 4.5 to 7.0%. Also, high-density lipoprotein cholesterol (HDL-C) is typically increased by 0.1-0.4 mmol/L and blood triglycerides (TAG) reduced by 0.10 mmol/L. RCTs published since 1999 have furthermore been reviewed with reference to detailed quality criteria in order to assess the magnitude of the effect reported in recent studies. Twenty-seven studies with 37 treatment arms a median intake of 25 g soy protein/day, and representing 2122 volunteers were identified and data were pooled to conduct a further meta-analysis specifically to address the extent of the effects of soyfoods on blood lipids post-1999. The results of this analysis indicate that LDL-C was reduced by 0.22 mmol/L - not dissimilar to the effects reported in other reported meta-analyses. In addition, HDL-C was increased significantly by 0.08 mmol/L - equivalent to an increase of 5%. The findings of the present analysis, and others published in the last decade, suggest that isoflavones *per se* are not instrumental in the lipid-altering effects of soyfoods. Plausible mechanisms that may responsible for the cholesterol reducing properties of soy protein will also be discussed. In addition areas for future research and public health implications will be highlighted.

In: Food Science Research Summaries. Volume 3
Editor: Lucille Monaco Cacioppo
ISBN: 978-1-63117-864-1

Chapter 76

SOY ISOFLAVONES IN BONE HEALTH AND FUNCTION

Alessandra Bitto*, Bruce P. Burnett and Francesco Squadrito
Department of Clinical and Experimental Medicine and Pharmacology,
School of Medicine, University of Messina, Messina, Italy

RESEARCH SUMMARY

Osteopenia and osteoporosis are metabolic conditions characterized by an imbalance between osteoclast relative to osteoblast activity. This metabolic balance in bone remodeling can be caused by either a deficiency in osteoblast activity (bone formation) or excessive osteoclast activity (bone resorption), leading to compromised bone architecture and increased risk of fracture. Epidemiological data suggest that intake of soy isoflavones confers a significant protective effect against the development of osteoporotic fracture in Asian populations.

There are considerable regional differences in soy protein and isoflavone consumption, with Asian populations consuming an estimated 25-50 mg of isoflavone aglycone equivalents daily, with 10% of the population consuming more than 100 mg isoflavones daily.

The relationship between isoflavone intake and bone loss is more robust in postmenopausal women, suggesting that isoflavone intake restores the metabolic balance of bone formation and resorption disrupted by menopausal ovarian hormone loss. In the United States and Europe soy consumption is only 5-10% of that in Asian countries, this justifies the search for isoflavone-based supplements, medical foods and drugs which can improve bone health, especially in postmenopausal women.

* Address correspondence and reprint requests to: Dr. Alessandra Bitto, MD, PhD, Assistant Professor of Pharmacology, Department of Clinical and Experimental Medicine and Pharmacology, Torre Biologica 5th floor, c/o AOU G. Martino, Via C. Valeria, Messina 98125, Italy, Phone: +39 0902213093 Fax: +39 0902213300, Email: abitto@unime.it

In: Food Science Research Summaries. Volume 3
Editor: Lucille Monaco Cacioppo

ISBN: 978-1-63117-864-1

Chapter 77

SOY'S POTENTIAL ROLE IN WEIGHT MANAGEMENT

Mark B. Cope, Ratna Mukherjea, Elaine Krul and David B. Allison
MBC, RM and EK (Solae LLC, St Louis, MO) and DBA,
University of Alabama at Birmingham, Birmingham, AL, US

RESEARCH SUMMARY

Globally, obesity and the related health consequences associated with excess body fat (heart disease and diabetes) continue to be a major public health problem. A healthy diet containing adequate protein may help people reduce energy intake, and during weight loss, protein can help preserve muscle mass that is often lost. Consumption of soy protein, a high quality plant-based protein, may offer several health benefits. Preclinical and clinical trials have been performed and the current data suggest that soy is as good as other protein sources for promoting weight loss and weight management and there is a suggestive body of evidence that soy may confer additional health benefits.

In this chapter the following topics related to soy and body weight will be addressed: 1) soy (protein) may help improve body weight and or fat loss while preserving lean mass as part of a healthy diet; and 2) soy (protein) may help improve risk factors related to glucoregulatory and cardiovascular health during weight loss. Soy, a vegetable-based source of high quality protein, has the potential to help individuals with body weight control.

In: Food Science Research Summaries. Volume 3
Editor: Lucille Monaco Cacioppo ISBN: 978-1-63117-864-1

Chapter 78

Antihypertensive Activity of Soy and Soy-Derived Compounds

Kathleen M. Eyster*[*]*, *Nikolai P. Breitkopf and Douglas S. Martin
Division of Basic Biomedical Sciences
Sanford School of Medicine of the University of South Dakota, Vermillion, SD, US

Research Summary

Studies of the effects of soy in laboratory animals routinely identify significant reductions in hypertension. Clinical studies in humans are less consistent in their documentation of blood pressure lowering effects of soy although a consensus is emerging that confirms an antihypertensive effect. The primary health benefits of soy are ascribed to soy isoflavones and soy protein. Antihypertensive mechanisms of soy isoflavones include increases in vasodilator activity and reduction in vasoconstrictor activity, inhibition of the renin-angiotensin system, antioxidant effects, and positive effects on renal function. Soy isoflavones are also known as phytoestrogens as they can activate the genomic estrogen receptor. Recent evidence indicates that isoflavones can also activate the peroxisome proliferator activated receptor (PPAR) as well as the pregnane X receptor (PXR) and that these genomic mechanisms may also be involved in the regulation of blood pressure. Soy protein has been shown to contain the bioactive peptide, lunasin, which may be involved in the antihypertensive effect of soy protein. These recent findings regarding PPAR, PXR, and lunasin, as well as reports that soy isoflavones and soy protein may exert synergistic effects on hypertension reduction, provide new excitement to the study of the antihypertensive effects of soy.

[*] Corresponding author: Phone: 605-677-5159, Fax: 605-677-6381, E-mail: Kathleen.Eyster@usd.edu

In: Food Science Research Summaries. Volume 3
Editor: Lucille Monaco Cacioppo
ISBN: 978-1-63117-864-1

Chapter 79

MECHANISMS OF ANTI-ATHEROSCLEROTIC ACTIVITY OF SOY AND SOY-DERIVED COMPOUNDS

Shanmugam Nagarajan*
Department of Pathology, Vascular Medicine Institute,
University of Pittsburgh Medical Center, Pittsburgh, PA, US

RESEARCH SUMMARY

Epidemiological evidence suggests that a high consumption of isoflavone-rich soy products is associated with a reduced incidence of cardiovascular disease (CVD). Moreover, soy-based diets have been reported to protect against the development of atherosclerosis in hyperlipidemic animal models. However, the underlying mechanism(s) for this protection remains unknown. Although atherosclerosis was traditionally considered a disease associated with impaired lipid metabolism, recent studies have convincingly delineated that uncontrolled oxidative stress and chronic inflammation is the principal contributing factor for the initiation and progression of atherosclerosis. Interaction between activated monocytes and vascular endothelial cells are an early event in atherogenesis.

There are two major components with potential bioactivity in soy diet or soy protein isolates: phytochemicals such as isoflavones associated with soy protein, and peptides generated from two of the major soy proteins such as β-conglycinin (or 7S globulins) and glycinin (or 11S globulins). Hence, in this review article the hypocholesterolemic and anti-inflammatory functions of soy-based diets contributing to the prevention of atherosclerosis are presented.

* Correspondene: Shanmugam Nagarajan, Ph.D. Department of Pathology, Vascular Medicine Institute, 200 Lothrop Street, BST E1526, University of Pittsburgh Medical Center, Pittsburgh, PA 15261, Tel: 412-648-8811, Email: NagarajanS@upmc.edu

In: Food Science Research Summaries. Volume 3
Editor: Lucille Monaco Cacioppo
ISBN: 978-1-63117-864-1

Chapter 80

EVALUATION OF EFFICACY OF SOY ISOFLAVONES IN REDUCING THE MENOPAUSAL SYMPTOMS

Márcia Constantino Colli,[1] Cinthia Gandolfi Boer,[2] Andréa Assunção Soares,[3] Niladê Neves Rosinski Rocha,[3] Adelar Bracht[3] and Rosane Marina Peralta[3,*]

[1]Departamento de Medicina
[2]Departamento de Análises Clínicas-Biomedicina
[3]Departamento de Bioquímica, Universidade Estadual de Maringá, Maringá – Parana, Brazil

RESEARCH SUMMARY

With the increase in life expectancy, many women live more than one third of their lives in menopause. Many of them show symptoms that affect negatively the quality of life and are exposed to a series of illnesses. The synthetic hormone replacement therapy (HT) is used during menopause to prevent symptoms and diseases related to menopause. Several alternative therapies have been studied for those patients suffering with menopausal symptoms and to whom the synthetic HT is contraindicated. The most known phytoestrogens are the soy isoflavones. However, studies on soy isoflavones consumed in different ways have shown inconclusive results regarding the improvement of some menopausal symptoms. The objective of this study was to assess the clinical efficacy, changes in the endometrium and in the vaginal epithelial maturation value of patients taking soy isoflavone to reduce the climacteric symptoms of menopausal women. Fifty one menopausal women were distributed in two study groups: group I received a daily dose of 100 mg of isoflavones (containing 70 mg of genistein) and group II received a daily dose of 100 mg of collagen (placebo group). The menopausal symptoms were assessed by the Kupperman index and by the intensity of hot flashes at the beginning and at the end of the 6 months of treatment. The effect on the endometrium was evaluated by determining its thickness through ultrasound and by vaginal

[*]Tel/Fax: +554430111372. E-mail adresses: rmperalta@uem.br; rosanemperalta@gmail.com

cytology at the beginning and end of the treatment. The results revealed that soy isoflavones were effective in reducing the menopausal symptoms evaluated by the Kupperman index (p=0.0047) and by the intensity of hot flashes (p=0.0010) and their use did not cause changes in the endometrium thickness and in the vaginal epithelial maturation value. This study suggests that the daily administration of 100 mg isoflavones was effective in reducing the clinical symptoms of menopause when evaluated by the Kupperman index and by the intensity of hot flashes.

In: Food Science Research Summaries. Volume 3
Editor: Lucille Monaco Cacioppo
ISBN: 978-1-63117-864-1

Chapter 81

SOY-DERIVED COMPOUNDS AND IMMUNE RESPONSE

Tohru Sakai, Emi Shuto and Rie Tsutsumi
Department of Public Health and Applied Nutrition
Institute of Health Bioscience
The University of Tokushima Graduate School, Tokushima, Japan

RESEARCH SUMMARY

The amount of soy products consumed in Asian countries is much greater than that in Western countries. Recent evidence indicates that soy and soy-derived components play a beneficial role in obesity, cancer, osteoporosis, and cardiovascular disease. The soybean isoflavone genistein is present at high levels in soy products. Genistein is structurally similar to 17β-estradiol, and genistein has been suggested to act as 17β-estradiol or an antagonist against 17β-estradiol. Genistein suppresses antigen (Ag)-specific immune response *in vivo* and lymphocyte proliferation response *in vitro*.

However, genistein enhances the cytotoxic response mediated by NK and cytotoxic T cells and the cytokine production from T cells. Thus, the effect of genistein on immunity is immune cell-dependent. Due to its unique effect on immune function, genistein has been used for treatment of diseases in animal models, and it has been found that genistein inhibits allergic inflammatory responses.

Equol is another class of soy isoflavones and exerts its effect strongly compared to the effect of other isoflavones. The immunomodulatory effect of equol is different from that of genistein. In this chapter, we summarize current studies related to the effect of the soy isoflavones on the immune system.

In: Food Science Research Summaries. Volume 3
Editor: Lucille Monaco Cacioppo
ISBN: 978-1-63117-864-1

Chapter 82

THE ROLE OF NON-ISOFLAVONE PHYTOCHEMICALS IN SOY-DERIVED HEALTH BENEFITS

Xianli Wu,* Jie Kang and Keshari M. Thakali
USDA Arkansas Children's Nutrition Center, University of Arkansas for Medical Sciences, Little Rock, AK, US

RESEARCH SUMMARY

Much attention has focused on researching the health benefits of soy. Based on many epidemiological and clinical studies, consumption of soy foods is associated with lower incidences of chronic diseases, including cardiovascular diseases, certain types of cancer and osteoporosis.

Over the last few decades, the focus has been to identify the bioactive components in soy and study their health benefits. Historically, isoflavones and soy protein were considered to be the major bioactive components in soy.

However, recent studies suggest that soy isoflavones or soy protein alone cannot explain all of the health benefits associated with soy consumption. There are many other potentially bioactive compounds such as soyasaponins, phytosterols and lignans that present in soybeans and soy-based foods. These non-isoflavone components may also contribute to the health benefits of soy, either by themselves or in combination with isoflavones and/or soy protein. These compounds vary in their chemical structures and their abundance in soybeans and soy-based foods.

The objectives of this chapter are to summarize the structural characterizations of non-isoflavone phytochemicals in soy and soy-based foods, and to outline the potential health effects of these phytochemicals.

*E-mail: wuxianli@uams.edu, Phone: 501-364-2813, Fax: 501-364-3161

In: Food Science Research Summaries. Volume 3
Editor: Lucille Monaco Cacioppo

ISBN: 978-1-63117-864-1

Chapter 83

EFFECT OF DIETARY FERMENTED SOYBEAN ON PHYSICAL ENDURANCE IN MICE

Nakamichi Watanabe and Satomi Sakuda
Department of Food Science and Nutrition
Showa Women's University, Tokyo, Japan

RESEARCH SUMMARY

Dietary soybean reported to increased physical endurance due to decrease plasma lipid peroxedation in 6th International Symposium on the Role of Soy, 2005. Intake of fermented soybean such as tempe expect to increase physical endurance more than that of non-fermented soybean, because we have been reported that the antioxidant activities of tempe was stronger than that of non-fermented soybean in the case of some assay such as balk system, emulsion system, red blood cells ghost system, and plasma system. In this study, we examined the effect of dietary non-fermented soybean and tempe on physical endurance and fatigue using swimming test in balb/c mice. Intake of non-fermented soybean decreased blood lactate levels and decreased plasma creatine kinase activities although swimming time was not affected at all. On the other hand, unfortunately, intake of tempe did not effect on physical endurance and fatigue although plasma lipid peroxidation decreased. In the mice fed non-fermented soybean, the plasma free fatty acid levels increased and glycogen contents in the skeletal muscle decreased although glycogen contents in the liver did not. These results suggested that the decreasing effect of dietary non-fermented soybean on fatigue might be induced by the improvement of utilization of glycogen in the skeletal muscle and fatty acid in the adipose tissue as energy source.

In: Food Science Research Summaries. Volume 3
Editor: Lucille Monaco Cacioppo

ISBN: 978-1-63117-864-1

Chapter 84

SOY CONSUMPTION AND SELF-REPORTED HEALTH

Eef Hogervorst,* Sumoryono, Ninu, Dwi Astuti, Sigit Prastowo, Amina Yesufu, Linda Kusdhany, Henry Setyawan, Dewi Praidini, Fidiansyah Murshid, Raden Ira Ismail, Philip Kreager, Xi Xin, Jen Stock, Angela Clifford, Stephan Bandelow, Sabarinah and Tri Budi Rahardjo

Respati University, Yogyakarta, SSEHS Loughborough University,
Center for Aging Studies, University of Indonesia

RESEARCH SUMMARY

Soy consumption is linked to both reduced- and increased-risk for age-related morbidity. This chapter covers our investigation of the association between soy consumption and self-reported health, as well as health-related factors including physician and hospital visits in the previous year. The study was carried out in three Indonesian cohorts on Java. These included rural Javanese and Sundanese elderly, as well as a mixed ethnicity urban cohort in Jakarta. Consumption of orange/red vegetables, fish and tempeh (a fermented soy product) was seen to have a protective association, while green vegetables were associated with poorer health, possibly due to the use of pesticides. Although higher levels of education were seen to have a protective association with better health, associations of foods with health were independent of education and occupation. This suggests that public health messages should focus on promoting consumption of non-animal based protein products and orange/red vegetables for better health. More research needs to focus on the association between poor health and consumption of products traditionally associated with better health, such as green vegetables.

*Tel: 01509 223020, Fax: 01509 223940, E-mail e.hogervorst@lboro.ac.uk

In: Food Science Research Summaries. Volume 3
Editor: Lucille Monaco Cacioppo
ISBN: 978-1-63117-864-1

Chapter 85

SOY ISOFLAVONES: BIOORGANIC AND MEDICINAL CHEMISTRY OF A POTENT COMPOUND CLASS

Sebastian Seibt
Organic Chemistry Laboratory, University of Bayreuth,
Bayreuth, Germany

RESEARCH SUMMARY

For ages soy has been a part of the vegetable diet of native peoples in East Asia because of its protective activity against a number of chronic diseases. Here, an overview of five abundant soy isoflavones Genistein, Daidzein, Glycitein, Biochanin A and Formononetin is provided. These simple phytoestrogens have revealed important biological activities and, thus, possess enormous health beneficial potential.

After a short introduction describing the history of their discovery, this chapter is dealing with bioorganic chemistry, spectroscopic investigations, synthetic chemistry and suitable labelling strategies of the five most important soy isoflavones, biosynthesis, and useful fermentation procedures.

In addition, the design, development and medicinal chemistry of potent new derivatives with improved biological activities compared with the parent soy isoflavones is described, followed by a short outlook on possible future developments in this important field of drug discovery. Pertinent literature is covered up to 2011.

In: Food Science Research Summaries. Volume 3
Editor: Lucille Monaco Cacioppo
ISBN: 978-1-63117-864-1

Chapter 86

SOY PLANT TISSUE CULTURE AS AN ALTERNATIVE FOR ISOFLAVONOID PRODUCTION

***A. Gueven*[1] *and Z. Hicsasmaz*[2]**
[1] Food Engineering Dept., Tunceli University, Tunceli, Turkey
[2] Food Engineering Dept., Trakya University, Edirne, Turkey

RESEARCH SUMMARY

Soybean (*Glycine max.*) is popular due to health consciousness, mad cow disease and high meat prices. Soybean consumption ranges between frozen green soybeans to cold cereals, non-dairy cheeses and desserts, milk and yogurt alternatives. Soybean is a source of edible oil, and the residual meal is used as a protein supplement for livestock. Soybean lipids are cholesterol-free. Soybean consists of 15 % dietary fiber and its protein content is between 36-46 %. Soy protein contains all essential amino acids. It has superior functional properties and is therefore a replacement for animal proteins in non-diary alternatives and meat analogs. Soybeans are the most abundant source of isoflavones. Isoflavones are anti-atherosclerotic; they decrease risk of coronary disease, improve body composition and prevent hormone-dependent cancers.

Plant cell cultures are attractive alternatives for secondary metabolite biosynthesis. Plant tissue cultures offer production of phytochemicals independent of geographical, seasonal and environmental factors. They ensure continuous production with uniform quality and yield. Cell growth in plant tissue cultures and secondary metabolite biosynthesis are correlated, but there also exist non-growth-related secondary metabolites. Chemical and physical stresses trigger biosynthesis of secondary metabolites. First, the best metabolic stress that triggers secondary metabolite biosynthesis is found. Then, optimum conditions needed in a bioreactor are investigated. Scale-up problems arise due to oxygen and nutrient transfer. Agitation to increase mass transfer rates cause death of the cells due to shear. Thus, airlift bioreactors are recommended for plant tissue cultures. The next handicap is the recovery of secondary metabolites, especially when they are intracellular. Effects of various chemical and physical stresses on cell growth and isoflavonoid biosynthesis by the soybean callus suspension culture have been investigated. Methyljasmonate was effective in triggering isoflavonoid biosynthesis. Pulsed electric field (PEF) is promising for bioreactor applications, since PEF not only performs as stress to induce biosynthesis, but it also is an agent of intracellular secondary metabolite excretion and recovery. Influence of biotic stress on isoflavonoid production by the soy plant cell culture has also been highlighted. Effect of immobilization on the growth of soy plant cell cultures and isoflavonoid biosynthesis has been studied on pilot scale bioreactors in conjunction with innovative engineering solutions for product recovery. However, efforts to extend results of small scale fermentations in flasks to a semi-continuous airlift bioreactor system were not very successful. Shear stress during aeration prevented the cells from staying in aggregate form. Detailed optimum engineering solutions are required to enable continuous production of secondary metabolites using plant cell cultures and will continue to be a challenging issue during the beginning of the twenty-first century.

In: Food Science Research Summaries. Volume 3
Editor: Lucille Monaco Cacioppo

ISBN: 978-1-63117-864-1

Chapter 87

REMOVAL OF ANTINUTRITIONAL FACTORS IN SOY-BASED FOODS USING LACTIC ACID BACTERIA

Jean Guy LeBlanc,[1,*] Marisa Garro,[1] Graciela Rollan[1] and Graciela Savoy de Giori[1,2]

[1]CERELA-CONICET, San Miguel de Tucuman, Argentina
[2]Cátedra de Microbiología Superior, Universidad Nacional de Tucumán (UNT), Tucumán, Argentina

RESEARCH SUMMARY

Soy products are excellent for their high protein content, and soy proteins contain enough of all the essential amino acids to meet biological requirements when consumed at the recommended level of protein intake. However, soybeans, as well as other legumes, characteristically contain high concentrations of antinutritional factors such as α-galactooligosaccharides (α-GOS) and phytates that can inhibit the absorption of many essential nutrients and cause serious physiological problems. Hydrolytic digestion of α-GOS is relatively weak in mammals because they do not possess α-galactosidase (α-Gal) in the upper gastrointestinal tract. The indigestibility of these soluble carbohydrates results in their delivery into the colon where they are rapidly fermented by the resident microbiota resulting in the production of large amounts of gas. This induced flatulence greatly hampers the acceptability of soy products as a major food source for humans and animals. The use of microbial α-Gal is a promising solution for the degradation of these undesirable α-GOS. Phytates, a common component of soybeans and grains can hinder the ability of humans (and other animals) to absorb crucial minerals such as calcium, magnesium, iron, and zinc. Certain microorganisms produce the enzyme phytase that degrades phytates and therefore may potentially improve mineral bioavailability and absorption. This chapter will discuss the use

[*]E-mail: leblanc@cerela.org.ar, jeanguyleblanc@hotmail.com

of lactic acid bacteria (LAB) that produce α-Gal and/or phytates in order to improve the nutritional value and acceptability of soy-based products.

In: Food Science Research Summaries. Volume 3
Editor: Lucille Monaco Cacioppo
ISBN: 978-1-63117-864-1

Chapter 88

Soy Ingredient Addition to Nutritionally Fortified Baked Goods

P. Roccia,[1] P. D. Ribotta,[2] G. T. Pérez[3] and A. E. León[4*]
[1]Universidad Nacional de Córdoba (UNC), Facultad de Ciencias Agropecuarias Instituto de Ciencia y Tecnología de Alimentos Córdoba (CONICET-UNC), Argentina
[2]Universidad Nacional de Córdoba (UNC), Facultad de Ciencias Exactas, Físicas y Naturales. Instituto de Ciencia y Tecnología de Alimentos Córdoba (CONICET-UNC), Argentina
[3]Universidad Nacional de Córdoba (UNC), Facultad de Ciencias Agropecuarias. Instituto de Ciencia y Tecnología de Alimentos Córdoba (CONICET-UNC), Argentina
[4]Universidad Nacional de Córdoba (UNC), Facultad de Ciencias Agropecuarias. Instituto de Ciencia y Tecnología de Alimentos Córdoba (CONICET-UNC), Argentina

Research Summary

Soy ingredients have been used to improve the nutritional profile of foods because of their high proportion of good quality proteins. The addition of soy ingredients (such as enzyme-active full-fat, heat-treated full-fat and enzyme-active defatted soy flours and commercially isolated soy proteins) to traditional foods can improve their protein quality. Despite nutritional value, the development of products with suitable physical and sensory properties has become necessary. It is known that components of soy ingredients interfere in gluten formation, weaken dough, reduce dough gas retention capacity and bread quality. Several studies suggest that the soybean protein state is an important factor in determining soy dough and bread properties. A better comprehension of the physicochemical and rheological changes in the soy/wheat protein and soy protein/starch system may lead to acceptable formulations. This chapter describes the effect of soy products on the rheological,

* P. Roccia: Avenida Valparaíso s/n - CC 509 - 5000 Córdoba – Argentina. P. D. Ribotta: Juan Filloy s/n - Ciudad Universitaria - 5000 Córdoba – Argentina. G. T. Pérez: Avenida Valparaíso s/n - CC 509 - 5000 Córdoba – Argentina. A. E. León: Avenida Valparaíso s/n - CC 509 - 5000 Córdoba – Argentina.

physicochemical and microstructure of the gluten network and their relationship to soy-wheat bread quality.

In: Food Science Research Summaries. Volume 3
Editor: Lucille Monaco Cacioppo
ISBN: 978-1-63117-864-1

Chapter 89

INCREASING THE BIOAVAILABILITY OF SOY ISOFLAVONES USING LACTIC ACID BACTERIA

Graciela Savoy de Giori,[1,2*] Jose Marazza,[1] Jean Guy LeBlanc[1] and Marisa Garro[1]
[1]CERELA-CONICET, San Miguel de Tucumán, Argentina
[2]Cátedra de Microbiología Superior, Universidad Nacional de Tucumán (UNT), Tucumán, Argentina

RESEARCH SUMMARY

Interest in soybeans and soy-based products has grown significantly in the last decade due to their reported nutritional and health-promoting benefits. Researchers have credited phytochemicals in soybeans, especially isoflavones, for some of these beneficial health effects. Soy isoflavones are reported to play a role in the prevention of osteoporosis, cardiovascular disease, and several hormone dependent cancers. Soybean and soy foods contain isoflavones which are found as a complex mixture of glucoside conjugates the most abundant being the glucosides of genistein and daidzein, known as genistin and daidzin. In the case of soymilk, isoflavones are present as either aglycones or glucosides depending on the soy source used in manufacture (raw soybean or soy protein isolate). The metabolic fate of soy isoflavones after consumption, as well as their biological activities depends on their chemical structure. In order to exert a biological effect, isoflavone glucosides are hydrolyzed by both intestinal mucosal and bacterial β-glycosidase releasing the aglycones. Several studies have demonstrated that lactic acid bacteria with β-glucosidase activity are able to increase the aglycone content during soymilk fermentation. This chapter will discuss the use of Lactic Acid Bacteria that produce β-glucosidase in order to improve the bioavailability of soy isoflavones.

---

* Corresponding author: Graciela Savoy de Giori. E-mail: gsavoy@cerela.org.ar.

In: Food Science Research Summaries. Volume 3
Editor: Lucille Monaco Cacioppo

ISBN: 978-1-63117-864-1

Chapter 90

COMBUSTION OF BIODIESEL FUEL MADE FROM SOYBEAN OIL

Charles K. Westbrook
Lawrence Livermore National Laboratory,
Livermore, CA, US

RESEARCH SUMMARY

Soybeans are an important source of oil that can be made into commercial quality diesel fuel and used to power automobiles and trucks. Soy biodiesel fuel is the dominant biodiesel fuel currently being used in the United States, but very little careful research has been conducted to determine the combustion properties of this transportation fuel. The present chapter describes the use of detailed chemical kinetic reaction mechanisms to study ignition and combustion of soy biodiesel fuels. These fuels are compared with other biodiesel fuels and with conventional diesel and automotive fuels derived from petroleum. The advantages and drawbacks of biodiesel fuels in general and soy biodiesel fuel in particular are described, and the principal elements of chemical kinetic modeling as a scientific discipline are presented in enough detail to be able to understand how soy biodiesel fuel compares with conventional and other transportation fuels.

In: Food Science Research Summaries. Volume 3
Editor: Lucille Monaco Cacioppo

ISBN: 978-1-63117-864-1

Chapter 91

SOY ISOFLAVONE GENISTEIN IN CANCER CHEMOPREVENTION: ROLE OF ENDOGENOUS COPPER AND INVOLVEMENT OF REACTIVE OXYGEN SPECIES

Haseeb Zubair,[1] Mohammad F. Ullah,[1,2*] Husain Y. Khan,[1] Fazlul H. Sarkar,[3] Aamir Ahmad,[1,3] and Sheikh M. Hadi[1]

[1]Department of Biochemistry, Faculty of Life Sciences, AMU, Aligarh, India
[2]Prince Fahad Research Chair, Department of Medical Laboratory Technology, Faculty of Applied Medical Sciences, University of Tabuk, Tabuk, KSA
[3]Department of Pathology, Karmanos Cancer Institute, Wayne State University School of Medicine, Detroit, MI, US

RESEARCH SUMMARY

Epidemiological studies have suggested that populations with high isoflavone intake through soy consumption have lower rates of breast, prostate and colon cancer. Isoflavone genistein in soybean is considered a potent chemopreventive agent against cancer. Although several mechanisms have been proposed, a clear mechanism of action of isoflavones against cancer is still not well understood. We have proposed a novel mechanism for the cytotoxic action of plant polyphenols; such as resveratrol, EGCG, gallic acid, as well as genistein; which may have implications in cancer chemoprevention.

We believe the mechanism of action of these polyphenols involves the mobilization of endogenous copper and subsequent generation of reactive oxygen species leading to DNA

* Corresponding author: Dr. Mohammad F. Ullah, Prince Fahad Research Chair, Department of Medical Laboratories Technology, Faculty of Applied Medical Sciences, University of Tabuk, Tabuk, KSA.

fragmentation and consequent cell death in cancer cells. The significance of this mechanism is enhanced in the light of the knowledge that copper levels are known to be considerably elevated in all types of cancer. In this chapter we first discuss isoflavone-induced DNA fragmentation in isolated lymphocytes. Such cellular DNA breakage is inhibited by copper specific chelators, but not by iron and zinc chelators, DNA breakage is also inhibited by scavengers of reactive oxygen species. Several studies, both *in vivo* and *in vitro*, have clearly shown that genistein is a promising therapeutic candidate for cancer chemoprevention as it is able to cause inhibition of human cancer growth via pathways leading to apoptotic cell death. We further discuss the ability of genistein to target endogenous copper in breast cancer cells leading to prooxidant signaling and consequent cell death. Recently we have shown that oral administration of copper in rats leads to elevated levels of copper in lymphocytes and plasma. Further the treatment of such lymphocytes with phytochemicals such as genistein, EGCG and resveratrol leads to a significant degree of DNA breakage.

In: Food Science Research Summaries. Volume 3
Editor: Lucille Monaco Cacioppo

ISBN: 978-1-63117-864-1

Chapter 92

CHEMOPREVENTIVE ROLE OF SOYBEAN PEPTIDES AGAINST CANCER

Chia-Chien Hsieh,[1] Blanca Hernández-Ledesma[2] and Ben O. de Lumen[1*]

[1]Department of Nutritional Sciences and Toxicology, University of California Berkeley, CA, US
[2]Institute of Food Science Research (CIAL, CSIC-UAM), Madrid, Spain

RESEARCH SUMMARY

Cancer is one of the most common causes of death in developed countries. Its incidence and mortality rates are increasing, thus cancer has become the medical challenge of this century. Accumulating evidence supports the vital role of lifestyle changes in cancer prevention and control. It has been estimated that at least 35% of all cancer deaths are associated with dietary factors. Epidemiological studies have linked soybean consumption with low incidences rates of an array of chronic diseases, including cardiovascular disorders and cancer. Over the past decades, a remarkable number of cell culture, animal experiments and human trials focused to support this association between soy consumption and lower cancer risk has been conducted. Among soybean compounds exerting chemopreventive properties, isoflavones and peptides are most extensively studied. This review summarizes the effects of soybean protease inhibitors and lectins against cancer, also mentions their demonstrated mechanisms of action. Special attention is given to peptide lunasin, discovered in our laboratory twelve years ago, whose verified properties make it a promising alternative for cancer prevention and/or therapy. Its chemopreventive activity revealed by *in vitro* and *in vivo* models, as well as its bioavailability and mechanism of action are included in this chapter.

* Corresponding author: Professor Ben O. de Lumen. 119 Morgan Hall, University of California Berkeley, CA, US, Phone: +1 510 642 8144 Ben O. de Lumen: nitto@berkeley.edu.

In: Food Science Research Summaries. Volume 3
Editor: Lucille Monaco Cacioppo
ISBN: 978-1-63117-864-1

Chapter 93

IMPACT OF SOYFOOD INTAKE ON BREAST AND ENDOMETRIAL CANCER RISK AND THYROID FUNCTION: EVALUATING SAFETY CONCERNS

Mark Messina*
Nutrition Matters, Inc., Port Townsend, WA, US

RESEARCH SUMMARY

Soyfoods have risen in popularity during the past 20 years in non-Asian countries in part because of consumer awareness of research suggesting that independent of their nutrient content, they may reduce risk of several chronic diseases. Many of the proposed benefits of soyfoods are attributed to their uniquely rich isoflavone content. Isoflavones are phytoestrogens although they are more accurately classified as selective estrogen receptor modulators. Despite proposed benefits, concern has arisen that the estrogen-like effects of isoflavones may cause untoward effects in certain individuals. In this review, background information on isoflavones is provided and then three safety concerns expressed about soyfoods are evaluated.

In vitro and in ovariectomized athymic mice, the primary soybean isoflavone genistein stimulates the growth of MCF-7 cells, an estrogen-receptor positive human breast cancer cell line. However, recent research raises questions about the utility of this model for providing insight into the effects of isoflavones in humans. Furthermore, the clinical data show that neither isoflavones from soyfoods nor supplements adversely affect markers of breast cancer risk and Chinese and American epidemiologic studies indicate post-diagnosis soy consumption improves the prognosis of breast cancer patients. Therefore, the evidence no longer justifies clinicians advising their patients against the consumption of soyfoods. The estrogen-like effects of isoflavones have also raised concern about the possibility that soyfoods could increase risk of endometrial cancer even though the limited epidemiologic evidence suggests the opposite may be the case. Most concerning, is a 5-year clinical study

* Corresponding author: Mark Messina: Nutrition Matters, Inc. 439 Calhoun Street, Port Townsend, WA 98368, E-mail: markjohnmessina@gmail.com.

that found isoflavone exposure slightly increased risk of developing endometrial hyperplasia. However, there were a number of limitations to this study and no other clinical data are supportive of these findings in that isoflavones have not increased endometrial thickness in studies that have been conducted for as long as 3 years. Finally, in vitro and in rats genistein inhibits the activity of thyroid peroxidase, an enzyme involved in the synthesis of thyroid hormones, and may also act as an alternate substrate to tyrosine for iodination. However, clinical studies show that long term exposure to high amounts of isoflavones does not adversely affect thyroid function in euthyroid individuals. On the other hand, preliminary data show that isoflavones may increase risk of subclinical hypothyroid patients progressing to overt hypothyroidism although this research also found that isoflavones favorably affected markers of diabetes and cardiovascular disease. Subclinical hypothyroid patients who begin to consume soy should have their thyroid hormone levels carefully monitored.

In: Food Science Research Summaries. Volume 3
Editor: Lucille Monaco Cacioppo
ISBN: 978-1-63117-864-1

Chapter 94

A Prospective Look at Genistein Use in Cancer Prevention and Treatment

Amanda C. Sehmer,[1] Aamir Ahmad,[1] Yiwei Li, Bin Bao[1] and Fazlul H. Sarkar[1,2*]
[1]Department of Pathology and [2]Oncology, Karmanos Cancer Institute, Wayne State University School of Medicine, Detroit, MI, US

Research Summary

Increased consumption of soy in Asian populations has been linked with reduced risk of several types of cancers when compared to high red meat consuming populations who have increased cancer risk. Therefore, the use of isoflavones found in the soybean has become the subject of much interest regarding their role in cancer prevention and therapy. Genistein is one such isoflavone and its anti-carcinogenic effects have been demonstrated in every step of cancer progression. Genistein inhibits proliferation and promotes apoptosis. Furthermore, its ability to inhibit metastasis of cancer cells through modulation of multiple signaling pathways, speaks not only for its capacity as a cancer preventative agent but also as a therapeutic agent. Genistein's inhibition of cancer cell proliferation and survival is in part due to its ability to inhibit several oncogenic pathways such as the Akt and NF-κB pathways, among others. Similarly, because chemotherapy and radiation therapy are associated with enrichment of drug resistant cells due to the activation of Akt and NF-κB pathways, which could be attenuated by genistein, genistein could ameliorate the unwanted effects of chemo-radio therapy. However, there are some controversies regarding the beneficial effects of genistein among different cancer types, which could in part be due to the structural similarities of genistein to estrogen. Moreover, emerging evidence suggest that the health benefit of genistein is indeed due to isoflavones found naturally and not due to pure genistein, which is supported by pre-clinical data in multiple cancer models and also supported by

* Corresponding author: Fazlul H. Sarkar, PhD, Department of Pathology, Karmanos Cancer Institute, 740 HWCRC Bldg, 4100 John R. Street, Detroit, MI 48201 USA, E-mail: fsarkar@med.wayne.edu, Tel: 313-576-8327, Fax: 313-576-8389.

limited clinical findings, suggesting that soy-derived isoflavone is likely to play an important role in the prevention and/or treatment of human malignancies.

In: Food Science Research Summaries. Volume 3
Editor: Lucille Monaco Cacioppo
ISBN: 978-1-63117-864-1

Chapter 95

METAL AND POLYCYCLIC AROMATIC HYDROCARBON CONTENT OF PRODUCTS IN SUGAR PRODUCTION

Zita Šereš,* Biljana Škrbić, Dragana Šoronja Simović, Biljana Pajin, Ljubica Dokić and Djordje Tadić
Faculty of Technology,
Department of Carbohydrate Food Engineering,
University in Novi Sad, Serbia

RESEARCH SUMMARY

Metal content in intermediate an in byproducts of sugar technology is important due to some of them are used in animal or in human nutrition. Numerous metals are necessary for the human organism, but in large amounts they are toxic. The metals, which are found in sugar beet, might exist in molasses and in dietary fibers from sugar beet and through it, they may be incorporated into human food. The aim of this work is to investigate the application of different sample preparation methods for determining the content of certain selected metals using atomic absorption spectrophotometer. It is determined the content of macro- and microelements in composite samples of sugar beet, dried sugar beet pulp, molasses and sugar. The division of each metal through the technological procedure of sugar production, from the row material through the intermediate products, to the final and byproducts is also determined. Experimental data proved that for determining metal content in sugar beet pulps both sample preparing methods could be used equally. Molasses has the highest content of potassium, about 37000 mg/kg, followed by sodium and calcium with 9000 and 3000 mg/kg, respectively. From original iron content in sugar beet there is appropriately only half of it left in dietary fibers (130 mg/kg - 180 mg/kg). From the analyzed microelements the most present in dried sugar beet pulp as in molasses is iron with average content about 150 mg/kg.

* E-mail: zitas@tf.uns.ac.rs.

Polycyclic aromatic hydrocarbons (PAHs) are fused-ring aromatic compounds, very important group of persistent organic pollutants present in the air, in water sources, in the soil, in sediments, in animals and plants in various industry products. Traces of these substances have been found in various food products. Most of them are very mutagenic and carcinogenic because of the ability of its metabolites to react with DNA inside the cell. Until industrial revolution, there was a balance between forming of PAH and natural degradation of PAHs. PAH, s has little but almost the same amount. With development of industry and intensive communication this is the misbalance in environment and the speed of PAHs deposition are increased. The mentioned substances are mainly formed during incomplete combustion of such are coal, gas and oil. By wet and dry deposition, they are maintained in all kind of matrix of living sources. The war in the Serbian Republic in 1999 destroyed large amounts of crude oil and derivates have consequence of possibility to carry PAHs into nutrition (food) chains.

Industry plant, such as sugar beet could be contaminated by PAHs by deposition from air and/or by transport and drying.

The aim of the technology of sugar-beet manufacturing is to produce sugar from row material under optimal technological and economic conditions. Among main product, sugar crystal, byproducts are molasses and sugar beet fibers. Sugar-beet fibers, mainly used as animal feed, are dried in filter dryer with rotational drum. Dryers could be heated by gases from their own heater, which are set for one kind of fuel or for heating could be used gases from their heating house, which passes through adequate plants to remove ash and soot. This is the reason why it is expected in dried sugar beet fibers the amount of PAHs to be high. Determination of PAHs in different sources from environment natural sources is complex and difficult. Difficult is due to PAHs are present in very little amounts and due to their present in complex matrixes which contains hundreds of different kind of organic compounds in different range of concentrations. Problems at identification are due to great number of PAHs izomers. Due to the above mentioned it is necessary to carry on the sample-preparing and the clean-up. The precision of the results is very influenced by sample preparing method.

In this work examines the presence of PAHs in sugar and in sugar beet fibers.

PAHs are examined by method which is recommended for this type of matrix, with some changes due to characteristics and complexity of samples. Beside PAH content in mentioned products of sugar factory, it is determined the reproductibility of the used method. For these experiments samples are used from two sugar factories from Vojvodina which are representative for given areas.

At sugar and sugar beet fibers samples preparing as extraction agents are used methanol, cyclohexane and dimethylformamide. The identification of PAHs is carried out using gas chromatography with flame ionization detector.

In: Food Science Research Summaries. Volume 3
Editor: Lucille Monaco Cacioppo
ISBN: 978-1-63117-864-1

Chapter 96

BIOPROTECTIVE EFFECTS OF SUCROSE AND TREHALOSE ON PROTEINS

Salvatore Magazù,[1,*] Federica Migliardo,[1,2] Antonio Benedetto,[1] Emanuele Calabrò,[1] Roberto La Torre[3] and Maria Teresa Caccamo[1]
[1]Dipartimento di Fisica e di Scienze della Terra,
Università di Messina, Messina, Italy
[2]Laboratoire de Dynamique et Structure des Matériaux
Moléculaires, Unesco-L'Oreal University of Lille I, France
[3]Department of Matter Physics and Electronic Engineering,
University of Messina, Messina, Italy

RESEARCH SUMMARY

The aim of the present chapter is to report some experimental findings on the dynamical effects induced by the addition of homologous disaccharides, i.e., sucrose and trehalose, to hydrated proteins. On purpose Elastic Incoherent Neutron Scattering (EINS) data, collected by the IN10 ILL spectrometer in the T=20 K ÷320 K temperature range, on dry and hydrated Lysozyme, also in the presence of disaccharides, are reported and discussed. Furthermore, Fourier Transform InfraRed (FTIR) spectroscopy data are reported in order to show the bioprotective effectiveness of trehalose connected with the application of an extremely low frequency electromagnetic field, at 180 μT.

In particular, the neutron scattering results highlight that the dynamical transition, registered at about T=220 K in hydrated Lysozyme, in the presence of disaccharides is shifted to a temperature of T=255 K; on the other hand, FTIR spectra show an increase of the CH_2 group spectral contributions at 1921 cm^{-1} and 1853 cm^{-1} after 3 hours of exposure, and that

* Corresponding author: Salvatore Magazù. Dipartimento di Fisica e di Scienze della Terra, Università di Messina, Viale F. Stagno D'Alcontres, n° 31, P.O. Box 55, S. Agata 98166 Messina, Italy. E-mail: smagazu@unime.it, tel. +39 0906765025 fax. +39 090395004.

such effect is not observed after the addition of disaccharides at the same exposure conditions.

In: Food Science Research Summaries. Volume 3
Editor: Lucille Monaco Cacioppo
ISBN: 978-1-63117-864-1

Chapter 97

INFLUENCE OF SUCROSE ON SURROUNDING WATER BY EXTENDED FREQUENCY RANGE DEPOLARIZED LIGHT SCATTERING

***L. Lupi*[1], *L. Comez*[1,2], *M. Paolantoni*[3], *P. Sassi*[3], *A. Morresi*[3] *and D. Fioretto*[1,4]**

[1]Dipartimento di Fisica, Università di Perugia, Perugia, Italy
[2]IOM-CNR c/o Dipartimento di Fisica, Università di Perugia, Perugia, Italy
[3]Dipartimento di Chimica, Università di Perugia, Perugia, Italy
[4]Centro di Eccellenza sui Materiali Innovativi Nanostrutturati (CEMIN), Università of Perugia, Perugia, Italy

RESEARCH SUMMARY

The role of disaccharides as bioprotectors is intimately connected with their interaction with hydration water. To investigate this issue, depolarized light scattering spectra of aqueous solutions of sucrose have been recorded in the range 0.3 GHz – 30 THz by combining the use of a double monochromator and a multipass Fabry-Perot interferometer. The spectra give us information on both the rotational diffusion of a single sucrose molecule and the effect induced by this solute on the picosecond dynamics of water. In particular, two distinct relaxation processes have been identified, and assigned to hydration and bulk water molecules.

The mobility of water localized within the sugar hydration shell is found to be lower than the mobility of water in bulk. The number of molecules involved in the slower dynamics and the extent of retardation are quantified and discussed in comparison with the results recently obtained for other mono and disaccharide solutions.

In: Food Science Research Summaries. Volume 3
Editor: Lucille Monaco Cacioppo
ISBN: 978-1-63117-864-1

Chapter 98

IMPROVEMENT OF COLOR AND FUNCTIONAL STABILITY OF STRAWBERRY JAM BY IMPREGNATION WITH SUCROSE

Yoshiyuki Watanabe*

Department of Biotechnology and Chemistry, Faculty of Engineering, Kinki University, Takaya, Higashi-Hiroshima, Japan

RESEARCH SUMMARY

Strawberry impregnated with sucrose was used for the jam preparation, and the effect of various impregnating conditions on the stability of anthocyanins in the jam was examined. The amount of sucrose in impregnated strawberry depended on the concentration of sucrose solution and the operation time for the impregnation. It was kinetically indicated that the impregnation of strawberry in advance of the jam preparation stabilized anthocyanins in the jam. The rate constants by the Weibull model for the color deterioration of the jam impregnated with sucrose and the degradation of anthocyanins in the jam were estimated, and the relationship between the two rate constants was observed to be almost linear.

The antioxidative ability of the strawberry jam correlated with the amount of anthocyanin in the jam. It was shown that the impregnation of strawberry with sucrose was effective for improvement of functional stability of strawberry jam.

* E-mail address: wysyk@hiro.kindai.ac.jp.

In: Food Science Research Summaries. Volume 3
Editor: Lucille Monaco Cacioppo
ISBN: 978-1-63117-864-1

Chapter 99

THE EFFECTS OF SUCROSE ON NEURONAL ACTIVITY

Elena Timofeeva and Arojit Mitra
Faculté de Médecine, Département Psychiatrie et Neurosciences,
Centre de recherche de l'Institut universitaire de cardiologie
et de pneumologie de Québec, Université Laval,
Québec (QC), Canada

RESEARCH SUMMARY

During the past few centuries humans have transformed from rear to voracious consumers of sugar. The present chapter sheds light on the questions of why we did increased our sugar consumption and what effects sucrose intake can produce on our mood and mind. Once consumed, sucrose sends signals to the brain via specialized taste receptors and glucosensing mechanisms. Sucrose intake activates in the brain the primary gustatory pathway and the brain reward system, which recognizes sweet taste of sucrose as rewarding. The role of the sweet taste system is to detect potentially high-energy sweet food that normally is relatively rare in the nature. The glucosensing mechanisms stimulate or inhibit food intake according to the energy needs. Because of the scarcity of sweet food in the nature, seeking this type of food requires some effort. Therefore, once such a food is found and consumed, this effort would be awarded by the brain reward system. These sophisticated systems work properly when their functions are not compromised by the unlimited availability of sweet palatable food, which may counteract the precise mechanisms controlling food intake. The experimental results obtained in animal models have shown that seeking and consumption of the regular non-sweet foods activate the medial hypothalamus which is directly related to the control of energy homeostasis and stress response. In contrast, seeking and consumption of sucrose inhibit the activity of the medial hypothalamus, but strongly activate the brain reward areas. This sucrose-induced shift in neuronal activity results in consumption of more and more palatable sweet foods at the expense of regular low-energy foods. Repeated overeating of sucrose can alter the neurochemical state of the brain reward regions, promoting sucrose craving and triggering sucrose binge-eating episodes. On the other hand, a sucrose-induced decrease in the activity of the medial hypothalamus directly affects the response of neuroendocrine system to stress. The stress-induced neuronal expression of stress neuropeptides as well as the release of plasma stress hormones is blunted by sucrose consumption. This 'anti-stress' effect of sucrose may explain overconsumption of sweets during stress even in the absence of hunger or needs for calories. This chapter reviews the neuronal effects of sucrose that from one side is beneficial by giving us energy and helping to relieve stress, or, from the other side, it may be detrimental by leading to uncontrolled craving for sucrose, eating disorders and obesity.

In: Food Science Research Summaries. Volume 3
Editor: Lucille Monaco Cacioppo
ISBN: 978-1-63117-864-1

Chapter 100

EFFECT OF THE ADDITION OF SUCROSE IN RHEOLOGICAL PROPERTIES

Priscila Anadão *

University of São Paulo –School of Engineering, Metallurgical and Material Science Department, Avenida Professor Mello Moraes,
São Paulo- SP, Brazil

RESEARCH SUMMARY

Sucrose is an important macronutrient that provides humans with a source of energy. Sucrose most important advantages include low cost and easy availability. Aiming to modify the rheological properties of a system, sucrose addition has been investigated and has shown to be successful in several areas. In the food area, sucrose is very important, as it can be used as an ingredient or a preserving agent and its effect in rheological properties is investigated in order to obtain, for example, information about differences in food texture. Sucrose has also been employed as an emulsifier in oil-water emulsions that reduces the interfacial tension, denoted by viscosity changes. The ceramic area has also used sucrose to adjust the forces between the particles that compose the dispersions since sucrose provides steric hindrance, hydration repulsion or electrosteric stabilization. Ergo, the objective of this chapter is to present literature examples regarding sucrose addition and its effect on the rheological properties of the mentioned areas.

* E-mail: priscila.anadao@gmail.com.

In: Food Science Research Summaries. Volume 3
Editor: Lucille Monaco Cacioppo
ISBN: 978-1-63117-864-1

Chapter 101

REACTIONS OF SUCROSE IN THE PRESENCE OF SOLID CATALYSTS

D. S. Pito and J. E. Castanheiro*
Centro de Química de Évora, Departamento de Química,
Universidade de Évora, Évora, Portugal

RESEARCH SUMMARY

Carbohydrates are the most abundant organic compounds on the world. Consequently, sucrose is a molecule abundant and inexpensive. The products of sucrose hydrolysis are glucose and fructose, which are used in the food industry. The sucrose derivatives, such as esters and ethers, have interesting properties due to the presence of lipophilic and hydrophilic moieties, provided by sucrose and alkyl chain. Traditionally, these reactions are carried out over homogenous catalysts, which are widely used in chemical processes to synthesize fine and bulk chemicals. However, they have serious drawbacks of hazards in handling, difficult separation and production of toxic waste. Solid catalysts can replace the homogeneous ones for simpler and environmentally more benign processes. Heterogeneous catalysts have been used in hydrolysis, esterification and etherification reactions of sucrose. In this work, the use of different solid catalysts for the conversion of sucrose (by hydrolysis, esterification and etherification) will be reviewed.

* Corresponding author. Tel.+351 266745311; Fax.: +351 266744971; E-mail address: jefc@uevora.pt.

In: Food Science Research Summaries. Volume 3
Editor: Lucille Monaco Cacioppo

ISBN: 978-1-63117-864-1

Chapter 102

CYTOTOXIC POTENTIAL OF SUCROSE OCTAACETATE FROM THE ETHANOL EXTRACT OF THE SEEDS OF *ANNONA CORNIFOLIA* A. ST.-HIL. (ANNONACEAE)

***Luciana A. R. S. Lima*[1,*], *Tânia M. A. Alves*[2], *Carlos L. Zani*[2], *Lúcia P. S. Pimenta*[3] and *Maria Amélia D. Boaventura*[3]**

[1]Campus Centro-Oeste Dona Lindu, Universidade Federal de São João Del-Rei, Divinópolis, MG, Brazil

[2]Laboratório de Química de Produtos Naturais, Centro de Pesquisas René Rachou/Fiocruz, Belo Horizonte, MG, Brazil

[3]Departamento de Química, Instituto de Ciências Exatas, Universidade Federal de Minas Gerais, Belo Horizonte, MG, Brazil

RESEARCH SUMMARY

Annona cornifolia A. St.-Hil. is a small annual perennial tree found in the Brazilian savannah and is popularly known as "araticum-mirim". The ripe orange fruit pulp is sweet and aromatic. The green fruit is popularly used in the treatment of ulcers. Sucrose octaacetate was obtained from the ethanol extract of *Annona cornifolia* seeds by acetylation with pyridine and acetic anhydride, and purified using column chromatography (silica gel and Sephadex LH-20). In a previous study, the antifungal activity of sucrose octaacetate was detected against the pathogenic fungus *Paracoccidioides brasiliensis*. *P. brasiliensis* isolates (Pb-18, Pb-11, Pb-01, Pb-B339, Pb-8 and virulent Pb-18) were susceptible to sucrose octaacetate. In the present work, the possible cytotoxic activity of sucrose octaacetate was evaluated by

* Correspondence: Prof. Dr. Luciana Alves Rodrigues dos Santos Lima - Campus Centro-Oeste Dona Lindu, Universidade Federal de São João Del-Rei, Rua Sebastião Gonçalves Coelho 400, Chanadour, Divinópolis, 35501-296, MG, Brazil. Tel.: +55 (37) 3221-1392. Fax.: +55 (37) 3221-1614. E-mail: luarsantos@ufsj.edu.br.

studying its effects against human tumor cell lines, including melanoma (UACC-62), renal carcinoma (TK-10) and breast cancer (MCF-7), using the sulforhodamine B assay. Sucrose octaacetate showed a cytostatic effect at 59.32, 63.98 and 32.51 μg/mL against the UACC-62, TK-10 and MCF-7 cell lines, respectively. In summary, this study provides evidence of the cytotoxic effect of sucrose octaacetate from the ethanol extract of *Annona cornifolia* seeds and also reports on the bioactivity of this compound for the first time. These results represent a step toward the use of this product as a drug candidate.

In: Food Science Research Summaries. Volume 3
Editor: Lucille Monaco Cacioppo
ISBN: 978-1-63117-864-1

Chapter 103

MYTHS AND FACTS ON CYTOSOLIC SUCROSE HYDROLYSIS

***Walter A. Vargas*[1,2,*,≠], *Mariana L. Martin*[1,†,#] and *Graciela L. Salerno*[1]**

[1]Instituto de Investigaciones en Biodiversidad y Biotecnología (INBIOTEC-CONICET) and Centro de Investigaciones Biológicas (FIBA), Argentina
[2]Centro Hispano-Luso de Investigaciones Agrarias (CIALE), Departamento de Microbiología y Genética, Universidad de Salamanca, Campus de Villamayor-Parque Científico, Spain

RESEARCH SUMMARY

Sucrose plays an essential role in Nature as part of the CO_2 assimilation pathway in the biosphere. In plants, sucrose is important for the allocation of carbon resources, and participates in a regulatory network that coordinates metabolism and development. Its metabolism is coordinated with the expression of photosynthesis-related genes by a feed-back regulation, and a sugar signaling pathway. Also, sucrose is accumulated in response to environmental stresses and its degradation has been associated with a high demand of carbon and energy in sink tissues.

In the last decades, several studies on sucrose degradation in plants brought to light novel trails on carbon trafficking and the metabolic paths leading to the regulation of gene expression. However, recent findings point to an intricate relationship between sucrose metabolism and the whole plant physiology. This review is focused on the elusive Alkaline/Neutral-Invertase (A/N-Inv) isoforms. These enzymes have been long believed as cytosol-exclusive proteins that hydrolyze the disaccharide for general or maintenance functions in plant cells. Nevertheless, the story has changed in the last few years when novel

[*] Corresponding author: Walter A. Vargas. E-mail: wvargas@usal.es.
[≠] These authors contributed equally to this work.
[†] E-mails: marimartin@fiba.org.ar, gsalerno@fiba.org.ar.

subcellular localizations and essential functions have been demonstrated for A/N-Invs. From being neglected as unimportant proteins, these enzymes became leading actors in sucrose catabolism with significant roles in root morphogenesis, photosynthetic apparatus biosynthesis, and abiotic stress adaptation.

The aim of this review article is to present a detailed and updated discussion of the most significant findings on the physiological properties of A/N-Invs. Also, we incorporate this group of invertases into the metabolic paths leading sucrose into the different circuits that govern carbon partitioning, storage and regulation of whole plant metabolism.

In: Food Science Research Summaries. Volume 3
Editor: Lucille Monaco Cacioppo
ISBN: 978-1-63117-864-1

Chapter 104

TEA POLYPHENOLS: HEALTH, BIOAVAILABILITY AND NOVEL USE

Kouass Sahbani Saloua[1,2,*]***, Kouass Salah***[3]
and Landoulsi Ahmed[1]
[1]Unité de recherche de Biochimie des lipides et interactions avec les macromolécules,.
Laboratoire de Biochimie et de Biologie moléculaire,
Faculté des sciences de Bizerte, Zarzouna, Tunisia
[2]Groupe en Sciences des Radiations, Département de Médecine Nucléaire et de Radiobiology, Faculté de Médecine, Université de Sherbrooke, Québec, Canada
[3]Laboratoire de Cristallochimie et Matériaux,
Faculté des Sciences de Tunis, El Manar II, Tunisia

RESEARCH SUMMARY

Tea, the dried leaves of the *Camellia sinensis* species of the Theaceae family, is a popular beverage with an annual production of three billion kilograms worldwide. Tea, brewed from the plant *Camellia sinensis* is consumed in different parts of the world as green, black or Oolong tea. Research findings suggest that the polyphenolic compounds, (−)-epigallocatechin-3-gallate found primarily in green tea, and theaflavin-3,3′-digallate, a major component of black tea, are the two most effective anti-cancer factors found in tea. The possible beneficial health effects of tea are being extensively investigated and have received a great deal of attention in recent times. Research conducted in many laboratory were indicate that tea can improve resistance against cancer, prevent cardiovascular diseases, promote digestive functions, stimulate the immune system, reduce inflammatory problems, and can have many other benefits. Tea catechins an important antioxidant that can be exploited as

* Author to whom correspondence should be addressed: Kouass Sahbani Saloua, Unité de recherche de Biochimie des lipides et interactions avec les macromolécules. 03/UR/0902. Laboratoire de Biochimie et de Biologie moléculaire Faculté des sciences de Bizerte Zarzouna7021, Tunisia. Groupe en Sciences des Radiations, Département de Médecine Nucléaire et de Radiobiology, Faculté de Médecine, Université de Sherbrooke, Québec, Canada J1H 5N4, Tel.: (001)819 346 1110 ext 12103 / (216) 23 347283, Fax: (001) 819-564-5442 /(216)-72-590566 E-mail: kouasssahbanisaloua@rocketmail.com.

dietary supplement in foods and beverages or as pharmaceutical/cosmetic use. However, the use of tea as a cancer chemopreventive or for other health benefits has been confounded by the low oral bioavailability of its active polyphenolic catechins, particularly epigallocatechin-3-gallate (EGCG), the most active catechin. The utilization of encapsulated polyphenols, instead of free compounds, can effectively alleviate these deficiencies. One approach to achieve this goal is the encapsulation of polyphenol with protein milk. It was found that use of thermally modified β-lactoglobulin conferred considerable protection to EGCG against oxidative degradation. Protein milk can increase the bioavailability of polyphenol also polyphenol can protected milk protein against oxidation. An addition of *Camellia sinensis* green tea extract before irradiation of caseins cow milk proteins was examined. We founded that the presence of *Camellia sinensis* green tea extract during irradiation in the presence of oxygen prevented the protein aggregation even at doses higher than 10 kGy, probably by scavenging oxygen radicals produced by irradiation. In this paper, we examine the available scientific information concerning tea, its polyphenol and human health. Also we report on studies that shed more light on the novel methods to increase the bioavailability of polyphenol and novel use of tea.

In: Food Science Research Summaries. Volume 3
Editor: Lucille Monaco Cacioppo
ISBN: 978-1-63117-864-1

Chapter 105

GREEN TEA AND ITS IMPACT ON HUMAN HEALTH: MYTH OR SCIENTIFIC TRUTH?

Branca M. Silva*
CICS-UBI - Centro de Investigação em Ciências da Saúde, Universidade da Beira Interior, Covilhã, Portugal; REQUIMTE/ Laboratório de Farmacognosia, Departamento de Ciências Químicas, Faculdade de Farmácia, Universidade do Porto, R. Aníbal Cunha, Porto, Portugal

RESEARCH SUMMARY

The origins of tea (*Camellia sinensis* (L.) O. Kuntze) are almost lost in time, with narratives that mix reality with many legends. For centuries, green tea has been widely used as food and medicine, since it provides, not only essential nutrients required for human life, but also other bioactive compounds for health promotion and disease prevention. In fact, some of these phytochemicals have served as sources of inspiration for generations of medicinal chemists, pharmacists, and nutritionists and will continue to provide humankind with valuable agents of potential use in research, prevention, and treatment of a number of chronic pathologies.

The chemistry of green tea is very complex, with thousands of phytochemicals, namely phenolic compounds, methylxanthines, organic acids, fibers, free amino acids, volatiles, between other classes of compounds. The potential bioactivities ascribed to this beverage include antioxidant, antimutagenic, anticarcinogenic, antimicrobial, anti-inflammatory, hypocholesterolemic, hypoglicemic and other clinically relevant activities.

Several scientific studies indicate that green tea is an excellent and inexpensive natural source of polyphenols, especially of catechins (flavan-3-ols). These compounds are well known for their strong antioxidant properties and are considered by many researchers as the

* CICS-UBI - Centro de Investigação em Ciências da Saúde, Universidade da Beira Interior, Av. Infante D. Henrique, 6200-506 Covilhã, Portugal; REQUIMTE/ Laboratório de Farmacognosia, Departamento de Ciências Químicas, Faculdade de Farmácia, Universidade do Porto, R. Aníbal Cunha, 164, 4050-047 Porto, Portugal, E mail: bmcms@ubi.pt.

main responsible for the beneficial health effects of this beverage and its derivatives (extracts).

In this chapter, I pretend to emphasize the virtues of this ancient drink in the light of the modern science. Thus, I will discuss briefly the aspects of its origin, cultivation and manufacturing process and will be focused in those related to its chemical composition and bioactivities, especially antioxidant and antitumor activities.

In: Food Science Research Summaries. Volume 3
Editor: Lucille Monaco Cacioppo

ISBN: 978-1-63117-864-1

Chapter 106

POTENTIAL HEALTH BENEFITS OF GREEN TEA: ROLE OF EPIGALLOCATECHIN-3-GALLATE IN PREVENTION OF INFLAMMATORY PATHOLOGIES

Esther Melgarejo, Miguel Ángel Medina, Francisca Sánchez-Jiménez and José Luis Urdiales*

Departamento de Biología Molecular y Bioquímica, Facultad de Ciencias, Universidad de Málaga, and CIBER de Enfermedades Raras, Málaga, Spain

RESEARCH SUMMARY

Tea is a popular beverage consumed worldwide that has been shown to have many health benefits. Its major polyphenols constituents, tea catechins, are the main components responsible for the medicinal effects of this beverage. Catechins exert vascular protective effects through multiple mechanisms, including antioxidative, anti-hypertensive, anti-inflammatory, anti-proliferative, anti-thrombogenic, and lipid lowering effects. Epigallocatechin-3-gallate (EGCG) is the most abundant catechin in green tea and presents potent anti-inflammatory effects by inhibiting enzyme activities and signal transduction pathways that play important roles in inflammation. After oral consumption, EGCG becomes bioavailable and several studies suggest that EGCG may directly interact with a large set of protein targets and alter the physiological response of the cells. This chapter highlights current developments about the effects of EGCG as an anti-inflammatory compound and its molecular targets that directly affect inflammation. These data support the use of EGCG as a possible chemopreventive agent with a potential to inhibit the development of multiple diseases, such as cardiovascular and neurodegenerative diseases and cancer.

* Corresponding author. E-Mail: jlurdial@uma.es.

In: Food Science Research Summaries. Volume 3
Editor: Lucille Monaco Cacioppo
ISBN: 978-1-63117-864-1

Chapter 107

EFFECTS OF GREEN TEA CATECHIN CONSUMPTION ON CARDIOVASCULAR DISEASES

***Jun-ichi Suzuki*[1,*], *Mitsuaki Isobe*[2], *Ryuichi Morishita*[3] *and Ryozo Nagai*[4]**
[1]Department of Advanced Clinical Science and Therapeutics, University of Tokyo, Japan
[2]Department of Cardiovascular Medicine, Tokyo Medical and Dental University, Japan
[3]Department of Clinical Gene Therapy, Osaka University,
[4]Department of Cardiovascular Medicine, University of Tokyo, Japan

RESEARCH SUMMARY

Green tea catechins are known to have many biological functions, including anti-inflammatory, anti-oxidative and anti-carcinogenic effects. These effects are induced by the suppression of several inflammatory factors including nuclear factor-kappa B (NF-κB), a multipotential promoter of matrix metalloproteinase (MMP), cytokines, and adhesion molecules. While these characteristics of green tea catechins have been well documented, the effects on inflammatory cardiovascular diseases have not yet been well investigated. In this article, we reviewed recent clinical and experimental data to reveal the anti-inflammatory effects of green tea catechins in cardiovascular diseases. We performed oral administration of green tea catechins into murine and rat models of cardiac transplantation, myocarditis and myocardial ischemia to clarify the effects on the inflammation-induced ventricular and arterial remodeling. From our results and other investigations, catechins are potent agents for the treatment and prevention of inflammatory cardiovascular diseases because they are critically involved in the suppression of proinflammatory signaling pathways.

* Correspondence to Jun-ichi Suzuki, Department of Advanced Clinical Science and Therapeutics, Graduate School of Medicine, University of Tokyo, 7-3-1 Hongo, Bunkyo-ku, Tokyo 113-8655, Japan; Phone 81-3-5800-9116, fax 81-3-5800-9182; E-mail : junichisuzuki-circ@umin.ac.jp.

In: Food Science Research Summaries. Volume 3
Editor: Lucille Monaco Cacioppo
ISBN: 978-1-63117-864-1

Chapter 108

ANTI-DIABETIC ACTIVITY OF BLACK, GREEN AND OOLONG TEA

Hardik Gandhi[1] and Aman Upaganlawar[2,*]
[1]Pharmacy Dept., Faculty of Tech. & Engg., the Maharaja Sayajirao University of Baroda, Vadodara, Gujarat, India
[2]SNJB'S SSDJ College of Pharmacy, Nashik, Maharashtra, India

RESEARCH SUMMARY

Black tea and green tea are the commonly known varieties of tea, used in different parts of the world. However, a third and lesser known variety, oolong tea is also available. Black tea is fully fermented; oolong tea is partially fermented, while green tea is not fermented at all. These differences in the processing techniques render these teas with distinctive chemical constitutions. Yet, various researchers have shown a positive role of each type of tea in ameliorating the knotty situation of diabetes. It has been published that consumption of tea, in any form, improves insulin resistance as well as glycemic control. Epidemiological studies have evaluated tea and its components and have shown a definite advantage of tea consumption in lowering the risks associated with diabetes. Moreover, the anti-hyperglycemic activity of tea has also been studied in vitro using the intestinal glucose uptake model. The role of tea polyphenols and catechins has been emphasized to a great extent in this regard. Apart from caffeine, catechins and polyphenols are found in tea in higher concentrations which are considered to be responsible for the anti-diabetic effects. Green tea was found to tender the renal complications of diabetes. Black tea was reported to reduce the risk of type 2 diabetes in Asian population. Oolong tea was found to augment the lowering of plasma glucose levels in combination with oral hypoglycemic agents. All these findings provide insights to the potential role of different varieties of teas in managing the difficult situation of diabetes.

* E mail: amanrx@yahoo.com.

In: Food Science Research Summaries. Volume 3
Editor: Lucille Monaco Cacioppo
ISBN: 978-1-63117-864-1

Chapter 109

MECHANISMS UNDERLYING ISCHEMIC STROKE PREVENTION BY TEA (*CAMELLIA SINENSIS*)

Yuri N. Clement*
Pharmacology Unit, Faculty of Medical Sciences, University of the West Indies, St. Augustine, Trinidad & Tobago

RESEARCH SUMMARY

Some prospective cohort and case-control studies suggest that long-term consumption of tea is associated with a reduced risk of mortality from ischemic stroke. Habitual tea consumption is also associated with a reduced risk of developing hypertension (an important cardiovascular risk factor for ischemic stroke). In stroke-prone spontaneously hypertensive rats, green tea consumption attenuated increases in blood pressure which corroborated with human observational studies. Tea flavonols enhance endothelial function by increasing cerebral blood flow and protecting against oxidative stress-induced ischemia. Green tea extract and its putative flavonol, epigallocatechin gallate (EGCG), significantly reduce infarct size in surgically-induced ischemic stroke in animal models. In vivo experiments show that EGCG, a potent free radical scavenger, activates endogenous antioxidant systems which reduce oxidative stress and ameliorate the sequel of events which lead to cerebral ischemia. EGCG reduces neuronal damage caused by chemically-induced oxidative stress, thus providing protection against ischemic stroke in in vitro models. Other studies have suggested that EGCG activates Nrf2 (a transcription factor which stimulates intracellular antioxidant signalling pathways) and increases the activity of heme oxygenase-I (a potent antioxidant enzyme). These mechanisms may explain, in part, the neuroprotective role of green tea against ischemic stroke. At the molecular level, components of green tea may provide neuroprotection by regulating various receptors, including $GABA_A$, and modulating intracellular signalling pathways, including STAT-1. The evidence from both pre-clinical and clinical studies suggests that tea may prevent ischemic stroke.

* E mail: Yuri.Clement@sta.uwi.edu.

In: Food Science Research Summaries. Volume 3
Editor: Lucille Monaco Cacioppo
ISBN: 978-1-63117-864-1

Chapter 110

Green Tea Polyphenolic Compounds and Cardiovascular Disorders

Aman Upaganlawar[1,*] ***and Hardik Gandhi***[2]
[1]Department of Pharmacology, SSDJ's College of Pharmacy, Neminagar, Chandwad. Dist: Nasik, India
[2]Department of Pharmacy, the Maharaja Sayajirao University, Vadodara, Baroda, India

Research Summary

Tea is one of the most popular drinks in the world, second only to water in terms of per capita consumption. Green tea is an extract of the leaves of the plant *Camellia sinensis* and it has been considered as a potential health beverage from ancient times in the system of medicine. Polyphenols are structurally diverse group of compound, ubiquitous in all plant kingdoms. As a result they form an integral part of human diet. Polyphenols are being widely used in the recent time as they possess antioxidant and free radical scavenging activities. Most of the polyphenols in green tea are flavanols, commonly known as catechins. The primary catechins in green tea are (-) epicatechin (EC), (-) epicatechin-3-gallate (ECG), (-) epigallocatechin (EGC), and (-) epigallocatechin-3-gallate (EGCG). In addition, caffeine, theobromine, theophylline and phenolic acids are also present as minor constituents. Epidemiological studies suggest that green tea and its catechins show varying health benefits in the prevention of cancer, diabetes, CNS disorders etc. The beneficial effects of green tea and its polyphenols in the prevention of cardiovascular disorders and their current status are reviewed in this chapter.

* Telephone: +91-9604033551; Email: amanrx@yahoo.com.

In: Food Science Research Summaries. Volume 3 ISBN: 978-1-63117-864-1
Editor: Lucille Monaco Cacioppo

Chapter 111

TEA AND HEALTHY SKIN

Neti Waranuch*[*] *and Wudtichai Wisuitiprot
Pharmaceutical Technology Dept, Faculty of Pharmaceutical Sciences,
Naresuan University, Thailand

RESEARCH SUMMARY

Tea is a well known beverage in the world due to its health benefits. A lot of studies indicated that tea contains more than 500 components that display several activities. Catechins are the most important component in a tea leaf. They present a great deal of antioxidant activity. Several reports revealed that catechins particularly epigallocatechin gallate (EGCG) are promising ingredients for protecting and nourishing skin. Catechins can scavenge free radicals and retard extracellular matrix degradation induced by UV radiation and pollution. Furthermore, catechins also directly affect the skin by activating collagen synthesis and inhibiting matrix metalloproteinase enzyme synthesis.

Based on the results in *in vitro* study, the benefits of tea looked promising. However, the results of the in vivo study have not been satisfactory. A few studies reported that tea and tea catechins significantly improved skin elasticity and increased dermis's thickness, therefore, skin wrinkle relief was obtained. The skin inflammation induced by UV radiation was decreased substantially when pre-treating the skin with tea catechins. On the contrary, a few studies revealed that topical formulation of tea catechins did not significantly improve skin health. The discrepancy in results with the in vivo study was due to the limitation of catechins's properties. Instability, less skin permeation and cutaneous metabolism were the crucial parameters in effectiveness of tea catechins. The health benefit from tea seems to be limited in real life, the appropriate preparations of tea application and delivery technology are the solution.

[*] Tel: +66 55 226577; Fax: +66 55 226577; E mail: netiw@nu.ac.th.

In: Food Science Research Summaries. Volume 3
Editor: Lucille Monaco Cacioppo
ISBN: 978-1-63117-864-1

Chapter 112

Epigallocatechin-3-gallate (EGCG), a Constituent of Green Tea, and its Antifibrotic Effect

O. A. Branford, A. O. Grobbelaar and K. J. Rolfe*
Institute of Plastic Surgery Research and Education, Department of Plastic Surgery, The Royal Free Hospital, Hampstead London, UK

Research Summary

Scarring and adhesions remain a major problem in medicine and surgery, causing unsightly repairs on the skin, disability if the scarring affects limbs or fingers and even death if scarring occurs in organs such as liver, kidneys or lungs. *Camellia sinensis* (green tea) is known to contain a rich source of poylphenols known as catechins. The major polyphenol in green tea is epigallocatechin-3-gallate (EGCG) which has been shown to have a wide biological activity. EGCG has been shown in *in vitro* and *in vivo* studies to reduce fibrosis and the contraction often associated with scarring. This review will summarise current scientific evidence of the role that EGCG plays in reducing scarring in a number of models. It will also review the potential biological processes that are altered with the administration of EGCG and its future potential as an anti-fibrotic therapy.

* E mail: work@kerstinrolfe.com.

In: Food Science Research Summaries. Volume 3
Editor: Lucille Monaco Cacioppo
ISBN: 978-1-63117-864-1

Chapter 113

TRANSLATIONAL RESEARCH OF THE GREEN TEA CATECHIN, EPIGALLOCATECHIN GALLATE, AS AN HIV-1 THERAPY

Christina L. Nance [*] ***and William T. Shearer***
Department of Pediatrics,
Baylor College of Medicine, Houston, TX, US

RESEARCH SUMMARY

Human immunodeficiency virus-1 (HIV-1) infection ultimately results in impaired specific immune function subsequent to the initial binding of the HIV-1-gp120, to the CD4 receptor. HIV-1 eludes the immune system and leads to the onset of acquired immunodeficiency syndrome (AIDS). Ever since the discovery of the virus as the causative agent, there has been an intense effort to develop therapeutic methods to inhibit or prevent infection. The current impact of the global AIDS epidemic is staggering with over 60 million people infected with HIV. Antiretroviral (ARV) therapy has decreased the number of AIDS cases dramatically. However, ARV therapy is not able to eradicate HIV-1 from patients completely. Furthermore, the toxicity of current available anti-HIV drugs makes it difficult to maintain patients' adherence to ARV therapy. The inevitable emergence of drug-resistant mutants, especially multi-drug resistant mutants, in response to ARV therapies will reduce the rates of success of ARV. There is an urgent need to expand the range of interventions to prevent HIV transmission and acquisition and development of novel anti-HIV agents is necessary.

One such candidate is the green tea catechin, epigallocathecin gallate (EGCG). In HIV-1 infection, EGCG is responsible for the reported antiviral effects of green tea. EGCG inhibits HIV-1 replication in human peripheral blood mononuclear cells (PBMC) in vitro with resultant decrease in HIV-1 p24 antigen concentration and reverse transcriptase.

[*] Corresponding author. Dept Allergy/Immunology, MC: FC330.01, Texas Children's Hospital, 6621 Fannin, Houston, TX 77030; Tel -1-832-824-2589; E mail: clnance@texaschildrens.org.

The translational research of this natural product from bench to bedside is chronicled here with evidence that the green tea component, EGCG, at physiologically relevant concentrations inhibits the binding of HIV-1-gp120 to CD4 over a broad spectrum of virus strains leading to the implementation of a Phase I/IIa clinical trial in HIV-1-infected individuals of a FDA-IND approved decaffeinated supplemental preparation of tea catechin, Polyphenon E, developed at NIH.

A major drawback of present conventional antiretroviral drugs is their high cost and unavailability in the Third World. A cheap, readily available, and safe treatment, such as could be provided by this complementary and alternative medicine, green tea catechin, would have an enormous impact on public health.

In: Food Science Research Summaries. Volume 3
Editor: Lucille Monaco Cacioppo
ISBN: 978-1-63117-864-1

Chapter 114

PASSIONFLOWER TEA CONSUMPTION AND SLEEP HEALTH

Alice Ngan and Russell Conduit*

School of Psychology & Psychiatry, Faculty of Medicine Nursing & Health Science,
Monash University, Melbourne, Victoria, Australia

RESEARCH SUMMARY

Herbal tea preparations have been traditionally used to remedy sleep disturbance for many years. Herbal plants such as valerian (*Valeriana officinalis*), kava kava (*Piper methysticum*), chamomile (*Matricaria recuita*), and passionflower (*Passiflora incarnata*), have long been used in sedative herbal teas. Although these herbs are widely added to commercial products with the prime purpose of remedying sleep, there is little or no scientific evidence for their efficacy. While such herbs are sold as teas, extracts and tablets, it is not clear whether herbal tea preparations contain any bioactive constituents that are effective beyond the relaxing effects of a warm beverage. This chapter describes the prevalence of mild sleep disturbance in the population, the lack of effective treatments that are readily available, and the therapeutic application of herbal medicines as alternatives or complements to pharmacological and psychological treatments. A review of the evidence regarding the effectiveness and potential side effects of sedative herbs, with an emphasis on *Passiflora incarnata* herbal tea, is then provided. Recent research suggests that *Passiflora incarnata* tea can promote better subjective sleep quality. Therefore, it seems that both widespread popular use and initial research evidence suggest that there is efficacy in the use of *Passiflora incarnata* herbal tea for treating mild fluctuations in sleep quality. Further scientific studies are needed to confirm recent findings, and to better understand the potential efficacy of *Passiflora incarnata* tea on the different aspects of sleep.

---

* E mail: russell.conduit@monash.edu.

In: Food Science Research Summaries. Volume 3
Editor: Lucille Monaco Cacioppo
ISBN: 978-1-63117-864-1

Chapter 115

Green Tea Consumption and Male Urinary Incontinence: A Community-Based Study in Japan

Andy H. Lee*
School of Public Health, Curtin Health Innovation Research Institute,
Curtin University, Perth, WA, Australia

Research Summary

Urinary incontinence (UI) is a distressing condition and costly problem in middle-aged and older people. A recent study suggested an inverse association between UI and habitual green tea consumption in middle-aged and older women. To investigate whether a similar relationship holds for male UI, seven hundred men aged 40-75 years were recruited from the community in middle and southern Japan. A validated food frequency questionnaire was administered face-to-face to obtain information on dietary intake and habitual tea consumption. UI status was ascertained using the International Consultation on Incontinence Questionnaire-Short Form. Among the 683 eligible participants, 49 men (7.2%) experienced urine leakage for the past 2.6 (SD 1.9) years. Their daily green tea intake (281 ml) was lower than others without the condition (311 ml), even though the prevalence of green tea drinking (88%) was similar between the two groups. Relative to non-drinkers, the adjusted odds ratios of UI were 0.93 (95% confidence interval 0.36 to 2.42) for drinking more than 700 ml and 0.75 (95% confidence interval 0.27 to 2.04) for drinking four or more cups daily. The corresponding dose-response relationships for these quantity and frequency of intake were not significant. The findings suggested no association between UI and habitual green tea consumption in middle-aged and older Japanese men.

* Corresponding author: Professor Andy H. Lee. School of Public Health. Curtin University, GPO Box U 1987, Perth, WA, 6845, Australia. Phone: +61 8 92664180; Fax: +61 8 92662958; Email: Andy.Lee@curtin.edu.au.

In: Food Science Research Summaries. Volume 3
Editor: Lucille Monaco Cacioppo

ISBN: 978-1-63117-864-1

Chapter 116

Greenhouse Tomato Yield and Solar Radiation

Tadahisa Higashide
National Agricultural and food Research Organization,
Institute of Vegetable and Tea Science, Japan

Research Summary

Tomato (*Solanum lycopersicum*) yield is determined by many processes such as leaf photosynthetic rate, fruit set, light interception, total dry matter production, distribution to fruits. These processes are called yield components. The important point is that the components are arranged in a hierarchy and that upper components are determined by related lower components, e.g., fresh fruit yield is directly determined by dry weight yield and fruit dry matter content. This chapter focused solar radiation and tomato yield components. Yield and yield components are affected directly or indirectly by solar radiation. Leaf photosynthetic rate affects directly on dry matter production per intercepted photosynthetic active radiation; light use efficiency (LUE). Light extinction coefficient; *k*, also affects on LUE. In a crop canopy that has a small *k*, light interception by lower leaves is higher than that in a large *k*. Although light-photosynthesis curve could limits leaf photosynthetic rate in upper leaves, the rate in the lower leaves in small *k* canopy is higher than that in large *k* canopy. Photosynthetic rate per plant is integrated leaf photosynthetic rate from lower to upper leaves. Thus, LUE in a crop canopy that has a small *k* is higher than that in a large *k*. Greenhouse tomato yield in The Netherlands has increased tremendously over the past 50 years. This increase in yield is caused by many factors, e.g., improvements in greenhouse transmissivity and developments in cultivation techniques such as longer cultivation, soilless culture, high wire training systems, CO_2 enrichment, environmental control by computers, and breeding for higher yielding cultivars. Higashide et al. investigated higher yielding cultivars of tomatoes that were released over the past 50 years. They reported that total dry matter production of plants increased with the year of release, dry matter partitioning to fruit of modern cultivars was not increased and fruit dry matter content was not decreased. Thus, an increase in yield over the past 50 years in Dutch tomato was caused by an increase in total dry matter production. Then, the increase in dry matter production was caused by high LUE. Light extinction coefficient in the crop canopy decreased while leaf photosynthetic rate increased significantly with the year of release.

In: Food Science Research Summaries. Volume 3
Editor: Lucille Monaco Cacioppo
ISBN: 978-1-63117-864-1

Chapter 117

TRANSLOCATION AND DISTRIBUTION OF PHOTOSYNTHATES IN TOMATO PLANTS

Takashi Nishizawa [*] ***and Yoshihiro Shishido***
Faculty of Agriculture, Yamagata University, Japan

RESEARCH SUMMARY

Carbon budgets of photosynthates in tomato plants as influenced by environmental factors were studied using radioisotopes as tracers. In $^{14}CO_2$-pulse feeding studies, both translocation and distribution of the ^{14}C-photosynthates were affected not only by the relative sink strength of each organ, but also by the phyllotaxis. Among the environmental factors, temperature was the most effective factor for altering the distribution rate of ^{14}C-photosynthates among sink organs but not so much for the export rate from the source leaf. Because the $^{14}CO_2$-pulse feeding method can detect only the destination of temporary assimilated carbon, the $^{14}CO_2$-steady state feeding (SSF) method was developed for clarifying the carbon budget for a whole day including the respiratory loss. As a result, translocation and distribution patterns of the ^{14}C-photosynthates by both $^{14}CO_2$-feeding methods were similar to each other, indicating that they were comparable techniques for detecting the destination of the assimilated carbon. In addition, the carbon budget of young tomato seedlings as affected by different environmental conditions, such as root temperatures, photoperiods and light intensities, were also investigated using the SSF method. Lastly, the positron-emitting tracer imaging system (PETIS) was introduced as a new tracing technology for visualizing the real-time movement of ^{11}C-photosynthates from the source leaf to sink organs.

---

[*] E-mail address: nisizawa@tds1.tr.yamagata-u.ac.jp.

In: Food Science Research Summaries. Volume 3
Editor: Lucille Monaco Cacioppo
ISBN: 978-1-63117-864-1

Chapter 118

Sensing, Control and Instrumentation during Tomato Growth

Federico Hahn*
Universidad Autonoma Chapingo, Estado de Mexico, Mexico

Research Summary

During tomato growth several variables should be controlled to make it profitable. In greenhouses the climate should be controlled to provide the adequate environment for the plants. Controlled irrigation and fertigation optimizes yield, saves water and fertilizers increasing producer's revenue. Radiation control for screen automation is essential to avoid tomato cracking and fruit water loss on hot summers. Integrated plant management detection was reviewed with trendy technology which can help growers to obtain better harvests. New humidity, radiation and electrical conductivity sensors are compared in base of their accuracy, range and output signals so producers can select the best option. Their application in climate and ferti-irrigation control with some up to date fertigation mechanisms show last technological advances. Implementation of closed loop fertigation can be economically implemented avoiding soil and water pollution.

* E-mail address: Irrigation Department, km 38.5 Carr. México - Texcoco, 56230, Chapingo, México; fhahn@correo.chapingo.mx.

In: Food Science Research Summaries. Volume 3
Editor: Lucille Monaco Cacioppo
ISBN: 978-1-63117-864-1

Chapter 119

LIBRARIES OF TOMATO MUTANTS INDUCED BY HEAVY-ION BEAM IRRADIATION: NEW EFFECTIVE METHOD FOR ANALYZING FUNCTIONAL GENOMICS

Shunsuke Imanishi*[*]*, Arisa Noguchi and Ichiro Honda
National Institute of Vegetable and Tea Science, National Agriculture and Food Research Organization, Japan

RESEARCH SUMMARY

Mutational approaches are highly instrumental for the study of the genetics and molecular bases of traits. Mutant plants are useful tools not only as parents of new cultivars but also as materials for assessing physiological mechanisms. To obtain information on these mechanisms, we induced mutations by irradiation with accelerated heavy ions in tomatoes. Irradiation with accelerated heavy ions, which was recently established as an effective method for inducing mutations in plants, could be a new approach for studying the functional genomics of tomato and those of the Solanaceae family.

[*] E-mail address: phppublish@ml.affrc.go.jp.

In: Food Science Research Summaries. Volume 3
Editor: Lucille Monaco Cacioppo
ISBN: 978-1-63117-864-1

Chapter 120

APPLICATION OF MOLECULAR METHODS FOR TOMATO BREEDING: THE SOLANUM/CLAVIBACTER PATHOSYSTEM

José Pablo Lara-Ávila[1], June Simpson[2] and Ángel Gabriel Alpuche-Solís[3]

[1] Laboratorio Nacional de Genómica para la Biodiversidad, Cinvestav Campus Irapuato, Guanajuato, México

[2] Departamento de Ingeniería Genética, Cinvestav Campus Irapuato, Guanajuato, México

[3] IPICYT, División de Biología Molecular, Instituto Potosino de Investigación Científica y Tecnológica A.C., SLP, México

RESEARCH SUMMARY

*Clavibacter michiganensi*s subsp. *michiganensis* (*Cmm*) is a Gram-positive phytopathogen causing bacterial wilt and canker of tomato (*Solanum lycopersicum*), an economically important disease producing economic losses worldwide. This chapter will review the current state of research on the interaction between *C. michiganensis* and *S. lycopersicum* with emphasis on the exploitation of molecular methods to study the genetic basis of resistance through gene expression profiles. A detailed example of gene expression profiling conducted at the early stages of bacterial wilt disease in tomato, using several resistant tomato-related wild species, including *S. peruvianum* LA2157, *S. peruvianum* LA2172, *S. habrochaites* LA2128, and a susceptible tomato species, to identify genes involved in the disease response will be presented. The potential of these findings and those from other recent studies for the identification of new candidate bacterial wilt disease resistance genes which could be subjected to further functional analyses to determine the molecular basis of incompatibility between wild species of tomato and *Cmm* will be discussed with the aim of providing a contribution to the development of more effective and sustainable *Cmm*-control methods.

In: Food Science Research Summaries. Volume 3
Editor: Lucille Monaco Cacioppo
ISBN: 978-1-63117-864-1

Chapter 121

BIOCONTROL OF FUNGAL DISEASES OF TOMATO: CONTRIBUTION OF *PSEUDOMONAS* SP. PCI2

Nicolás A. Pastor, Susana B. Rosas, Javier A. Andrés, Marina C. Niederhauser and Marisa Rovera
Laboratorio de Interacción Microorganismo–Planta. Facultad de Ciencias Exactas, Físico-Químicas y Naturales, Universidad Nacional de Río Cuarto. Río Cuarto, Córdoba, Argentina

RESEARCH SUMMARY

Tomato is the most diffused vegetable plant worldwide. Thus, demand for tomato cultivation, production and commerce is continually increasing. Production of tomato fruits is destined for direct consumption and for processing into industrialized products, such as puree, pulp, juice and sauces.

Fungal diseases are one of the major factors limiting tomato production. These diseases are managed by the application of chemical fungicides. However, chemical fungicides are costly, can cause adverse environmental effects, can result in health hazards to humans and other organisms, and can induce pathogen resistance. Currently, it is widely accepted that biological control of plant diseases, involving the use of antagonistic microorganisms, is an alternative or a supplemental way to reduce the use of chemicals in agriculture for disease management.

A large number of microorganisms from the rhizosphere have been considered to play a key role in sustainable agriculture because of their biocontrol potentials and plant growth-promoting activities. Among these, the soil-borne fluorescent *Pseudomonas* have received particular attention because of their catabolic versatility, their excellent root-colonising abilities, their adaptation to environmental stresses and their capacity to aggressively compete with other microorganisms. The identification of antagonistic strains represents only the first step towards the development of effective biological control strategies. After the identification of organisms as potential antagonists, it is important to determine the specific mechanisms, interactions and conditions responsible for biological control. The understanding of these characteristics will allow scientists to establish both the limitations and the full potential of biocontrol for developing effective strategies for disease management.

Tomato is susceptible, among other fungi, to *Phytophthora infestans*, the causal agent of late blight, *Fusarium oxysporum* f.sp. *lycopersici* (Fol), which causes a severe wilt disease, *F. oxysporum* f.sp. *radicis-lycopersici* (Forl), the causal agent of crown and root rot, *Alternaria solani*, the causal agent of early blight, *Verticillium dahliae* and *V. albo-atrum*, which cause vascular wilt, and to *Sclerotium rolfsii* and *Alternaria alternata*. These two later fungi cause heavy losses in the yield of tomato. *S. rolfsii* is a major problem for the cultivation of tomato in the tropical regions of the world, causing damping-off in nursery seedlings as well as stem rot, wilting and blight in adult plants. *A. alternata* is frequently isolated from diseased tomato plants and causes stem canker and leaf spot.

In this chapter, we will review some studies on biological control of tomato fungal diseases by fluorescent *Pseudomonas*. We will also contribute data from our own work on the antagonistic activity of strain *Pseudomonas* sp. PCI2, isolated from tomato rhizosphere, against *S. rolfsii* and *A. alternata*.

In: Food Science Research Summaries. Volume 3
Editor: Lucille Monaco Cacioppo
ISBN: 978-1-63117-864-1

Chapter 122

BEGOMOVIRUS DISEASE: THE MOST DEVASTATING DISEASE OF CULTIVATED TOMATO

Walter N. Leke[1*], Abwe N. M. Dione[2], Sali A. Ndindeng[1], and Noé Woin[1]

[1]Institute of Agricultural Research for Development (IRAD), Yaounde, CR, Cameroon

[2]Institute of Agricultural Research for Development (IRAD), Ekona, SWR Cameroon

RESEARCH SUMMARY

Recently, several begomoviruses (genus *Begomovirus*; family *Geminiviridae*) have been associated with leaf curl and yellow leaf curl symptoms in tomato crops worldwide. These viruses are transmitted by the whitefly *Bemisia tabaci* (Genn.) sibling species group, and they have a genome of single stranded (ss) DNA (about 2.7-2.8 kb) with an organization typical for Old World monopartite begomoviruses. In addition, many monopartite begomoviruses are associated with alphasatellite and betasatellite molecules that may influence symptom development, pathogenicity, and virulence. Wild, uncultivated plant species are commonly infected with viruses, and may serve as reservoirs of viruses also pathogenic to cultivated plants, and can exert dynamic forces on the ecology of natural systems.

* Corresponding author: Email: lekwat@yahoo.com.

In: Food Science Research Summaries. Volume 3
Editor: Lucille Monaco Cacioppo
ISBN: 978-1-63117-864-1

Chapter 123

Tomato Antioxidant Content as Influenced by Conventional and Organic Production Methods and Post Harvest Management

Ruplal Choudhary*[*] and S. Alan Walters
Department of Plant, Soil and Agricultural Systems, Southern Illinois University, Carbondale, Illinois, US

Research Summary

Consumers are becoming more aware of the nutritional benefits that vegetable provide to their diets and are interested in purchasing those that have increased nutrient or antioxidant content. Many studies have focused on the nutritional value of vegetables produced in organic and conventional (or non-organic) production systems; and, the nutritional value of organic vegetables compared to those produced using conventional production methods is widely debated. However, many consumers often perceive organically produced vegetables as being safer (reduced pesticide residues), produced using more environmental friendly production methods, and more nutritious compared to those produced using conventional production practices. This chapter will focus on the comparison of organic and conventional production systems for the production of two antioxidants in tomato fruit, ascorbic acid and lycopene. In addition, a brief review of post-harvest management system for tomato packaging, transportation and storage will be discussed in relation to antioxidant content of fruit.

[*] E-mail address: choudhry@siu.edu.

In: Food Science Research Summaries. Volume 3
Editor: Lucille Monaco Cacioppo
ISBN: 978-1-63117-864-1

Chapter 124

SUSTAINABLE INTEGRATED DISEASE AND PEST MANAGEMENT IN COMMERCIAL TOMATO PRODUCTION IN OPEN FIELDS

***Bombiti Nzanza*[1]*, Kgabo Pofu*[2] *and Phatu Mashela*[3*]**
[1]Head of Department of Research, ZZ2 Bertie van Zyl, Mooketsi
[2]Researcher, Department of Nematology and Entomology, Agricultural Research Council, Rustenburg
[3]Professor of Plant Protection, University of Limpopo, Sovenga, South Africa

RESEARCH SUMMARY

A set of research-based environment-friendly approaches are being adopted and applied by tomato (*Solanum lycopersicum*) farmers to respond to increased threats of below- and above-ground virulent pathogen, nematode, mite and insect pests. Primarily, soil health is improved through the application of disease-suppressive and nutrient-enriched compost, cover cropping, effective microorganisms and minimum tillage tactics. Plant-parasitic nematodes are managed using fermented crude extracts of locally available plant materials. Nursery seedlings are inoculated with *Trichoderma harzianum* and arbuscular mycorrhizal fungi for improving tolerance to soil-borne diseases and enhancing overall crop quality and development. Crop protection is achieved through the application of plant extracts, while tomato fields are enriched through augmentation of beneficial insects, coupled with limited application of synthetic chemical pesticides when necessary. A case study for this model is a large-scale commercial tomato farming operation under open fields in South Africa, which has gradually adopted *natuurboerdery* or a hybrid of conventional and natural farming approaches in order to strike a balance between chemical farming and organic farming, with the aim of enhancing sustainable tomato production.

* E-mail: phatu.mashela@ul.ac.za.

In: Food Science Research Summaries. Volume 3
Editor: Lucille Monaco Cacioppo
ISBN: 978-1-63117-864-1

Chapter 125

TRENDS IN FOREIGN TRADE COMPETITIVENESS FOR FRESH TOMATO

Jaime de Pablo,[1] Miguel Giancinti[2] and Juan Uribe[3]
[1]University of Almería, Spain
[2]Office MAG, Argentina
[3]University of Almería, Spain

RESEARCH SUMMARY

Drawing on Porter's competitive forces, we can say that the rivalry of direct competitors is greatest when there is greater concentration, a slow growth market, fixed costs and high stocks, differentiated products, and no absence of rules. Given this situation, the circumstances are perfect for a risk of entry by potential competitors and intelligence systems must be used to reduce uncertainty. It also decreases when there are high barriers to entry or strong retaliatory measures. Possible barriers can be achieved through economies of scale, product differentiation, increased investment, access to distribution channels and government policies. The threat of product substitution is another variable that affects direct competition. In the case of tomatoes, other vegetables can be used in salads. The bargaining power of suppliers and customers depends on relative concentration, producer/client related quality between the product and raw material, product differentiation and the distribution of added value in the final product. This article analyzes the tomato from the perspective of competitiveness in production and consumption in 2004-2010.Thisstudy will analyze the most relevant trends and forecast the main trends for the coming years. The article will first show a structural analysis of the major export countries, after which it will follow the major consumers and trade.

In: Food Science Research Summaries. Volume 3
Editor: Lucille Monaco Cacioppo
ISBN: 978-1-63117-864-1

Chapter 126

THE CHEMICAL COMPOSITION AND NUTRITIONAL VALUE OF TOMATOES

M. Hernández Suárez, E. M. Rodríguez Rodríguez and C. Díaz Romero*

[1] Agriculture Technology National Centre "Extremadura" CTAEX, Ctra. Villafranco-Balboa km 1.2, Villafranco del Guadiana, Badajoz, Spain
[2] Analytical Chemistry, Nutrition and Food Science Department. La Laguna University, Av. Astrofísico Francisco Sánchez s/n, La laguna, Santa Cruz de Tenerife, Spain
[3] Analytical Chemistry, Nutrition and Food Science Department. La Laguna University, Av. Astrofísico Francisco Sánchez s/n, La laguna, Santa Cruz de Tenerife, Spain

RESEARCH SUMMARY

The tomato has traditionally been classified as a *dessert vegetable*. However, at present, this view has shifted because of a new concept which claims that the tomato is a source of interesting nutrients and bioactive components, especially antioxidants, which have positive effects on human health. Several parameters such as soluble solids, sugars, acidity and pH are important to establish the quality of both fresh market and processing tomatoes. The main components responsible for tomato taste and aroma associated to the quality are sugars, organic acids, free amino acids, salts and antioxidants. Oxalic acid, ascorbic acid and lycopene have interesting effects on human health. As with other vegetable foods, tomatoes have an adequate mineral and trace element composition to benefit and maintain human health.

The scientific community is focused on identifying the cause and the action mechanism which increase the concentrations of these biologically active compounds. The tomato, during ripening, undergoes a wide range of biosynthetic as well as hydrolysis reactions that dramatically affect the final fruit composition. Ascorbic acid and some phenolics tend to be accumulated from the green to mid-ripe stage, whereas the total carotenes constantly increase

* E-Mail address: cdiaz@ull.es.

their concentrations during the ripening process. Furthermore, under water and salt stress, tomatoes mature earlier and accumulate more lycopene in the period before harvest. The chemical composition depends on the simultaneous interaction of genetic aspects, environmental parameters and cultural practices. Multivariate analysis is a suitable tool to show the complex interactions between chemical composition and factors of production.

In: Food Science Research Summaries. Volume 3
Editor: Lucille Monaco Cacioppo ISBN: 978-1-63117-864-1

Chapter 127

FACTORS AFFECTING THE ANTIOXIDANT CONTENT OF FRESH TOMATOES AND THEIR PROCESSED PRODUCTS

Norma Francenia Santos-Sánchez, Rogelio Valadez-Blanco and Raúl Salas-Coronado*
Instituto de Agroindustrias, Universidad Tecnológica de la Mixteca,
Huajuapan de León, Oaxaca, México

RESEARCH SUMMARY

Worldwide, *Solanum lycopersicum* (tomato) is the most important vegetable, mostly due to its sensorial properties and also because it is a rich source of carbohydrates, fiber and antioxidants. In recent years there has been a growing interest in the study of antioxidants in vegetables due to the effects of these compounds on human health. The antioxidant activity of tomato fruits is attributed to vitamin C, phenolic compounds and carotenoids, particularly lycopene. This latter compound is of essential importance in tomato´s definition as a functional food, given its proven properties to fight prostate cancer. The antioxidants in tomato fruits have different degrees of liability in terms of exposure to air, storage and processing temperatures and enzymatic activity in fresh tomato. The degradation of these compounds in tomato fruits and its products is synergistic. Also, it has been observed that the concentration of antioxidants in fresh tomato depends on the variety and growing conditions. In processed tomato fruits, the antioxidant content is additionally influenced by the sample preparation (juice, puree, slices, etc.), processing temperature and time, and extent of water removal. Different methods have been developed to reduce the degradation of antioxidants due to processing. These methods include the use of coadjuvants, such as maltodextrins in juice spray drying, sodium metabisulfite in tray drying of tomato slices, etc. Therefore, the study of the antioxidant content in the fruit from fresh tomato to its final products is a topic of paramount interest to the farming and food processing industries; however, it has not been

* E-mail address: rsalas@mixteco.utm.mx. Phone: +52(953)5320214 ext.400.

comprehensively addressed in the literature. Thus, this chapter describes and analyzes the factors affecting the antioxidant content in fresh tomatoes and its processed products.

In: Food Science Research Summaries. Volume 3
Editor: Lucille Monaco Cacioppo
ISBN: 978-1-63117-864-1

Chapter 128

LYCOPENE IN TOMATO FRUITS AND ITS CONTENT VARIATION DURING THE RIPENING PROCESS AFTER HARVEST

Francelina Costa[1], Maria de Lurdes Baeta[2], David Saraiva[2], Maria da Conceição Castilho[2], Manuel Teixeira Verissimo[3] and Fernando Ramos[2*]

[1]Centro de Saúde de Torres Novas, R. Dr. Pedro GM Salazar, Torres Novas, Portugal

[2]CEF – Centro de Estudos Farmacêuticos da Universidade de Coimbra Health Sciences Campus, Pharmacy Faculty, Coimbra University, Azinhaga de Santa Comba, Coimbra, Portugal

[3]Serviço de Medicina Interna dos Hospitais da Universidade de Coimbra, Faculdade de Medicina Universidade de Coimbra, Coimbra, Portugal

RESEARCH SUMMARY

Lycopene is a carotenoid with no pro-vitamin A activity. However, lycopene is a very powerful singlet oxygen scavenger with an important protective effect against oxidative damage that also acts as an immune system stimulator. Among other benefits, lycopene intake appears to reduce the incidence of several cancers, including prostate, stomach and lung cancers and also plays a role in the prevention of certain cardiovascular diseases.

Thus, in this chapter, a review of the properties of lycopene with respect to human health was described, and a discussion of the lycopene content in tomato fruits during ripening after harvest was conducted on the basis of a comparative study.

The lycopene contents were determined in three types of tomato fruits (round, elongated and cherry). The extraction of the lycopene was achieved with a mixture of hexane, acetone and ethanol. An aliquot of the organic layer was evaporated to dryness. The dry residue was

* Corresponding author: Phone number: +(351) 239 488492, Fax number: +(351) 239 488503, Email address: fjramos@ci.uc.pt

recovered in the mobile phase and injected into an HPLC-UV system with detection at λ = 450 nm. The results revealed that the lycopene content varies after harvest throughout the ripening process. A significant increase in the lycopene content occurs at the maturation stage of turning tomatoes. Although lycopene synthesis occurs after harvest, the concentrations of lycopene in green tomatoes were still consistently below those in red tomatoes. However, an increase in the lycopene content was not observed after the harvest of red tomatoes. A maturation period greater than eleven days after the harvest consistently led to a decrease in the total lycopene content.

In: Food Science Research Summaries. Volume 3
Editor: Lucille Monaco Cacioppo

ISBN: 978-1-63117-864-1

Chapter 129

HEALTH PROPERTIES OF TOMATO ANTIOXIDANTS

Giulia Graziani[1,*], Antonello Santini[1] and Alberto Ritieni[2]
[1]Department of Food Science, University of Napoli "Federico II", Portici, Napoli, Italy
[2]Department of Pharmaceutical and Toxicological Chemistry, University of Napoli "Federico II", Napoli, Italy

RESEARCH SUMMARY

Epidemiologic evidence of a protective role for fruits and vegetables in cancer prevention is substantial. Antioxidants are important in disease prevention in both plants and animals, inhibiting or delaying the oxidation of biomolecules by preventing the initiation or propagation of oxidising chain reactions. Tomato is a versatile vegetable that is consumed fresh as well as in the form of processed products. Lycopene, the primary carotenoid in tomatoes, is frequently hypothesized to be one of the principle phytochemicals in tomato products with anticancer properties. Although lycopene is the most abundant carotenoid in tomato products, the tomato is also an excellent source of other nutrients and phytochemicals that may also contribute to the health benefits of tomatoes. Other tomato components such as phenolics, flavonoids and vitamins C and E are also responsible for the antioxidant capacity of raw tomatoes and processed tomato products. Epidemiological studies regarding intake of tomatoes and tomato-based products and blood lycopene reported an inverse associations between tomato intake or blood lycopene level and the risk of cancer. The evidence for a benefit was strongest for tumors of the prostate, lung, and stomach. Numerous other potentially beneficial compounds are present in tomatoes, and, conceivably, complex interactions among multiple components may contribute to the anticancer properties of tomatoes. Antioxidant nutrients are believed to slow the progression of atherosclerosis because of their ability to inhibit the tissue damage deriving to oxidative process. Several studies reported that thermal processing enhanced the nutritional value of tomatoes by

* Corresponding author: Giulia Graziani, Department of Food Science, University of Napoli "Federico II", Via Università 100, 80055 Portici (Napoli), Italy. Tel/Fax.: 0039 (81) 2539317. E-mail address: giulia.graziani@unina.it.

increasing the bioaccessible lycopene content and total antioxidant activity and are against the notion that processed fruits and vegetables have lower nutritional value than fresh produce. Other studies demonstrated that lycopene is rather thermally stable in tomato-based food systems during mild thermal treatments, while more intense thermal treatments can cause lycopene degradation reactions. Tomato, one of the most important vegetables worldwide, contains a range of flavonoids and phenolic acids in addition to lycopene, which are regarded as potentially useful compounds with respect to health benefits. Composition data in fresh tomatoes vary due to genetic and environmental factors and cultural practices. Breeding programs aim to produce tomatoes with enhanced levels of flavonoids and other phenolics.

In: Food Science Research Summaries. Volume 3
Editor: Lucille Monaco Cacioppo
ISBN: 978-1-63117-864-1

Chapter 130

The Protective Effect of Tomato Juice on the Respiratory System

G. S. Maritz and M. Mutemwa
Department of Medical Biosciences, University of the Western Cape, Bellville, South Africa

Research Summary

Tomato fruits are rich in various nutrients such as vitamins and minerals. It is also a source of fibre. Tomato fruits are a rich source of anti-oxidants and phytonutrients, such as vitamin C, lycopene, carotenes, vitamin E, polyphenols and anthocyanin. The anti-oxidants present in dietary tomato products, such as tomato juice, are associated with improved human health. There is also evidence of protection against prostate cancer. It was also demonstrated that tomato juice protects the adult human against the damage induced by tobacco smoke. Maternal nicotine intake during pregnancy and lactation prevents the programming of the lungs against premature aging and the development of emphysema later in life. Studies showed that if the mother's diet is supplemented with tomato juice, it prevents the decrease in the anti-oxidant capacity of the pregnant and lactating mother as well as that of the fetus and neonate and thus prevents the nicotine-induced programming. Consequently the protection of the baby by the mother is improved. This ensures that the respiratory system of the baby develop normal even when exposed to high levels of oxidants and nicotine via the placenta and breast milk. As a consequence the susceptibility of the individual to develop respiratory disease after birth and later in life will be reduced compared to those not exposed to the anti-oxidants of tomato fruits and thus not protected against oxidants that are inhaled or present in the food that the mother ate.

In: Food Science Research Summaries. Volume 3
Editor: Lucille Monaco Cacioppo
ISBN: 978-1-63117-864-1

Chapter 131

Trace Elements: Effect on Tomato Plant and on Human Health After Consumption of Tomato Fruit and Tomato Fruit Food Products

Charalampos Proestos,[1,*] Ioannis N. Pasias,[2] Vassiliki Papageorgiou,[3] Konstantinos Barmperis[3] and Nikolaos S. Thomaidis[2]

[1]National and Kapodistrian University of Athens, Department of Chemistry, Food Chemistry Laboratory, Panepistimiopolis Zografou, Athens, Greece
[2]National and Kapodistrian University of Athens, Department of Chemistry, Laboratory of Analytical Chemistry, Panepistimiopolis Zografou, Athens, Greece
[3]Hellenic Food Authority (EFET), Athens, Greece

Research Summary

The determination of trace elements in tomato fruits and tomato fruit food products has become very important in the last few years since it gives information about the contamination and helps to increase canned food quality and safety. A risk of contamination of the food chain may arise when heavy metals accumulate in plant tissues at concentrations above the threshold levels believed to threaten the health of consumers. Lead, cadmium, arsenic and mercury are the metals and metalloids with the highest toxic potential among those present in foods. Other metals including copper, iron, tin, zinc, antimony, aluminium, manganese, and chromium, which are present in trace amounts in the human diet, may also be the source of toxic hazards to humans under specific conditions. Moreover, the content of these elements in tomato food products varies depending on the type of product and on the procedure followed for the preparation and the type of package material. This chapter

* E-mail: harpro@chem.uoa.gr.

deals with the risk assessment of trace elements found in tomato fruits and tomato fruit food products on human health, the effect of trace elements on tomato plants and the analytical methodologies followed for their accurate and precise determination. The risk assessment of some toxic elements is also discussed.

In: Food Science Research Summaries. Volume 3
Editor: Lucille Monaco Cacioppo

ISBN: 978-1-63117-864-1

Chapter 132

NUTRACEUTICALS RECOVERY FROM TOMATO PROCESSING WASTE AND BY-PRODUCTS: LYCOPENE

Antonello Santini[1*], Giulia Graziani[1] and Alberto Ritieni[2]
[1]Department of Food Science, University of Napoli "Federico II," Portici (Napoli), Italy
[2]Department of Pharmaceutical and Toxicological Chemistry, University of Napoli "Federico II," Napoli, Italy

RESEARCH SUMMARY

Nutraceuticals are food or food products that can have health and medical benefits, e.g., the prevention and treatment of diseases. Among these, carotenoids, have a special interest for their high activity as natural antioxidants compounds. These molecules can play a relevant physiological role, and are contained in many fruits and vegetables like tomatoes. There is a growing interest worldwide by many companies in performing the extraction of carotenoids (lycopene, beta-carotene, etc.) from tomato wastes (peels and seeds) in order to meet the requirements and need of the international nutraceuticals, food additives, pharmaceutical, and cosmetic markets. The health benefits of these dietary antioxidants and their use in nutrition to supplement and correct dietary needs, has increased the demand for fruit and vegetable products with recognized and improved antioxidant quality. Tomato (*Solanum Lycopersicon*) is particularly rich in these molecules that are only partially recovered and available after the industrial processing. Lycopene can be an example of the versatility of the carotenoids. It is often included among the major components of pharmaceuticals for the treatment of many pathologies, and it is widely used in cosmetics formulations for its high solubility in fats and oils. It is worth to note that carotenoids like lycopene are still among the major components of waste originating from tomato processing plants, and are dumped with the waste, considered as low value by-product and useless material to discard. The amount of tomato by-products deriving from industrial processes is growing annually. In Europe it is estimated that about

* Corresponding author: Antonello Santini, Department of Food Science, University of Napoli "Federico II", Via Università 100, 80055 Portici (Napoli), Italy. Tel/Fax.: 0039 (81) 2539317, E-mail address: asantini@unina.it.

ten million tons of tomatoes are processed by the food industry, and the deriving wastes can be quantified in about 0.1 million tons. These residues can be used as a source of carotenoids and mainly of lycopene, reincorporated in low-quality tomato products, and also used as an ingredient addressed to animal foodstuff. Countries insisting on the Mediterranean Sea, e.g., Italy, are the main producers of tomatoes in Europe, and the waste and residues of the industrial processing of tomatoes represent a low cost and relatively cheap source of molecules with nutraceutical potential use.